Ben Zion Kon

University Mathematics
Handbook

Ben Zion Kon

University Mathematics

Handbook

Senior Editors & Producers: Contento
Translated by: Zvi Hazanov
Editor: Sherrill Layton
Graphic Design: Michael Gorelik
Cover Design: Benjie Herskowitz

ISBN-13: 978-152-298-994-3

ISBN-10: 152-298-994-3

International sole distributor: Contento
22 Isserles Street, 6701457, Tel Aviv, Israel
Netanel@contento-publishing.com
www.ContentoNow.com

Ben Zion Kon

University Mathematics

Handbook

- ♦ Vectors and analytic geometry
- ♦ Single-variable and multivariable differential and integral calculus
- ♦ Vector analysis
- ♦ Complex functions
- ♦ Series and Fourier Series
- ♦ Ordinary and partial differential equations
- ♦ Fourier and Laplace Transforms

To my wife Adela

Table of contents

Introduction

This book aims to guide university students enrolled in various mathematics courses of all levels.

It will help the readers refresh their knowledge, complementing partial knowledge of certain subjects, and finding the necessary formulas to solve various problems.

It contains definitions, formulas, and theorems for most mathematical courses given at the Technion, Israel Institute of Technology, as well as other higher education institutions.

For students of all levels, the book is most useful for helping students with their home assignments and preparing for exams.

The material divides into fifteen parts, dealing with the following subjects:

- Vectors and analytic geometry
- Single-variable and multivariable differential and integral calculus
- Vector analysis
- Algebra
- Complex functions
- Ordinary and partial differential equations
- Fourier series
- Fourier and Laplace transform.

Definitions, Theorems, Rules and Formulas accompanied by examples can be found by reference to the detailed table of contents in the front of the book or to the extensive index in the back of the book.

22

I hope the readers find this book most useful.

I wish to thank Dr. Michael Gorodetzki for his useful comments.

I also wish to thank Ms. Roni Mor for so diligently printing and Ms. Iris Weinstock for the book's graphic design.

Ben Zion Kon

I. Basic Concepts

- Integers set: $\mathbb{Z} = \{\ldots -3, -2, -1, 0, 1, 2, \ldots\}$

- Natural numbers set: The positive integers $\mathbb{N} = \{1, 2, 3, \ldots\}$

- Rational numbers set: $\mathbb{Q} = \left\{ \dfrac{m}{n} : m, n \in \mathbb{Z}, n \neq 0 \right\}$

- Real numbers set: $\mathbb{R} = \{x : -\infty < x < \infty\}$

- Irrational numbers set: $\mathbb{R} - \mathbb{Q}$ (all real numbers that are not rational)

- Complex numbers set: $\mathbb{C} = \{a + bi : a, b \in \mathbb{R}, i^2 = -1\}$

- Infinity: ∞

- An open interval: $(a, b) = \{x : a < x < b, x \in \mathbb{R}\}$

- A closed interval: $[a, b] = \{x : a \leq x \leq b, x \in \mathbb{R}\}$

- A half-closed interval: $[a, b) = \{x : a \leq x < b, x \in \mathbb{R}\}$ or $(a, b] = \{x : a < x \leq b, x \in \mathbb{R}\}$

- A ray: $(-\infty, a] = \{x : x \leq a, x \in \mathbb{R}\}$, $(a, \infty) = \{x : x > a, x \in \mathbb{R}\}$, $(-\infty, a)$ $[a, \infty)$

- The ε neighborhood of x_0 is the interval $(x_0 - \varepsilon, x_0 + \varepsilon)$: $(\varepsilon > 0)$

- $A \Rightarrow B$: Proposition A leads to proposition B, or proposition B is derived from proposition A. For example: $x > 2 \Rightarrow x^2 > 4$

- $A \Leftrightarrow B$: Propositions A and B are equivalent. For example: $|x| > 2 \Leftrightarrow x^2 > 4$

- $\forall$: "For all." For example: $\forall n \in \mathbb{N}$ "for all natural n "

- $\exists$: "There exists." For example: $\forall x, \exists y : x + y = 0$ "for all x, there exists y, such that $x + y = 0$ "

- $\in$: "Is a member of." For example: If a is a member of set A, then $a \in A$

- $e = 2.71828\ldots$

- $\pi = 3.14159\ldots$

- (n factorial) $n! = 1 \cdot 2 \cdot 3 \cdot \ldots \cdot n$, $0! = 1$

- Stirling's formula: $n! \approx \dfrac{n^n}{e^n} \sqrt{2\pi n}$ as $n \to \infty$

- $\dfrac{n!}{(n-k)! k!} = C_n^k = \begin{pmatrix} n \\ k \end{pmatrix}$, $0 \le k < n$, $n \in \mathbb{N}$

- $C_n^k = C_n^{n-k}$

0.1   Areas

a. The area of a triangle:

1. $S = \dfrac{1}{2} a \cdot h_a$, where h_a is the altitude of side a .

2. Heron's formula: $S = \sqrt{p(p-a)(p-b)(p-c)}$ where a,b,c are the triangle sides and $p = \dfrac{1}{2}(a+b+c)$ is half of the perimeter.

3. $S = p \cdot r$, where p is half of the perimeter and r is the radius of the circle in the triangle.

4. $S = \frac{1}{2} a \cdot b \cdot \sin \alpha$, where α is the angle between sides a and b.

b. The area of a parallelogram is the product of the side's length and the height of that side.

c. The area of a rhombus is half the product of the lengths of its diagonals.

d. The area of a rectangle is the product of its height and its width.

e. The area of the trapezoid is equal to half the product of the length of an altitude and the sum of the lengths of the bases.

f. The area of a circle is $S = \pi r^2$, where r is the radius of the circle.

g. The lateral surface area of a right circular cone is $S = \pi r \ell$, where r is the radius of the circle and ℓ is the lateral altitude of the cone (see next Figure).

h. The lateral surface area of a ball with an R radius is $S = 4\pi R^2$.

0.2   Volumes

a. The volume of a cuboid with a,b,c dimensions is $V = a \cdot b \cdot c$.

b. The volume of a parallelepiped is the product of the area of its base (S) and the altitude of its base (h): $V = S \cdot h$.

c. The volume of a pyramid is a third of the area of the base multiplied by the altitude of the base: $V = \frac{1}{3} S \cdot h$.

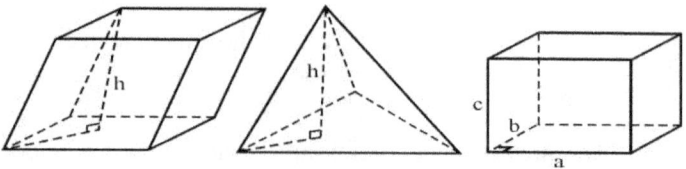

d. The volume of a right circular cylinder is $V = \pi r^2 h$, where r is the radius of the circle and h is the altitude of the cylinder.

e. The volume of a right, circular cone is $V = \frac{1}{3}\pi r^2 h$, where r is the radius of the base and h is the altitude of the cone.

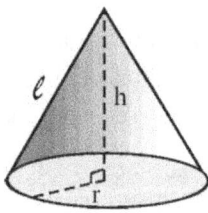

f. The volume of a ball is $V = \frac{4}{3}\pi R^3$, where R is the radius of the ball.

0.3   Greek Alphabet

α	alpha	ν	nu
β	beta	ξ	xi
γ	gamma	o	omicron
δ	delta	π	pi
ε	epsilon	ρ	rho
ζ	zeta	σ	sigma
η	eta	τ	tau
θ	theta	υ	upsilon
ι	iota	φ,φ	phi
κ	kappa	χ	chi
λ	lambda	ψ	psi
μ	mu	ω	omega

II. Functions

Chapter 1
General Properties of Functions

1.1   Definition

Let D and E be two sets of real numbers.

A function is a given rule fitting each number x in set D, a one and only one number y of set E.

That function is denoted as $y = f(x)$, which says f of x, or $f: D \to E$, which reads f is D into E.

Set D is called the function's domain.

x is called an input variable or argument.

If $y = f(x)$, then y is said to be the **image** of x, or x is the **preimage** of y.

If $f(x)$ is described by an algebraic expression, then the maximum set of real numbers for which the expression exits is **the natural domain** of $f(x)$.

$f(x)$ is **bounded** on domain D if there exists a number M such that $|f(x)| \le M$ for all $x \in D$.

Example: The domain of $y = \sqrt{x^2 - 4}$ is $x \le -2$, $x \ge 2$.

The graph of function $y = f(x)$ is a set of all pairs of coordinates $(x, f(x))$ where x belongs to D.

1.2   Increasing and Decreasing Functions

The function $y = f(x)$ **is increasing** in domain E if for every two points x_1 and x_2 from E, and where $x_1 < x_2$ there holds $f(x_1) \le f(x_2)$.

The function $y = f(x)$ **is decreasing** in domain E if for every two points x_1 and x_2, which are members of E, and where $x_1 < x_2$ there holds $f(x_1) \ge f(x_2)$.

If the inequality is strict, then the function is **strictly increasing** (or decreasing).

Example: The function $y = x^2$ is increasing in the domain $(0, \infty)$ and decreasing in $(-\infty, 0)$, and is neither increasing nor decreasing in any interval containing the point $x = 0$.

The graph of an increasing function in an interval is a curve ascending from left to right.
The graph of a decreasing function in an interval is a curve descending from left to right.

1.3   Odd and Even Functions

$f(x)$ is called an **even function** if for all x in its domain there is also $-x$ in its domain, such that $f(x) = f(-x)$.

Example: $y = \sqrt{x^2 - 4}$, $y = \dfrac{1}{x^2} + 1$, and $y = \cos x$ are even functions.

$g(x)$ is called an **odd function**, if for all x in its domain there is also $-x$ in its domain, such that $g(x) = -g(-x)$.

Example: $y = \sqrt[3]{x^5 + x}$, $y = \sin x$, and $y = \tan x$ are odd functions.

Example: $y = x^3 + 1$ is neither an even nor an odd function.

The sum of two even (or odd) functions is an even (or odd) function.

The product of two even (or odd) functions is an even function.

The product of an even function and an odd function is an odd function.

A graph of an even function is symmetric about the y axis.

A graph of an odd function is symmetric about the origin.

1.4   Periodic Function

$f : D \rightarrow R$ is called a **periodic function** if there exists a positive number T such that for all $x \in D$ there is also $x \pm T \in D$ and there holds $f(x \pm T) = f(x)$.

The minimum T, if it exists, is called the **period** of $f(x)$.

Example: $y = \sin x$, and $y = \cos x$ are periodic functions with a 2π period.

1.5   One-to-One Correspondence Function (Bijective Function)

$f : D \rightarrow E$ is a one-to-one correspondence function $(1 \leftrightarrow 1)$ if for all y of E there exists at most one x of D such that $y = f(x)$.

In other words, $f : D \rightarrow E$ is a one-to-one correspondence function if for all x_1, x_2 of D if $f(x_1) = f(x_2)$ then $x_1 = x_2$.

Example: $y = x^2$ is a bijective function in the domain $D = \{x : x > 0\}$ and not bijective in $\mathbb{R}$, since for $x_1 = -2$, $x_2 = 2$ there holds $f(-2) = f(2) = 4$.

1.6   Surjective Function

The function $f : D \to E$ is a **surjective function** if for all $y \in E$ there exists a $x \in D$ such that $y = f(x)$.

Examples:

a. $y = \sin x : R \to [-1, 1]$ is a surjective but not one-to-one correspondence function.

b. $y = \cos x : [0, \pi] \to [-1, 1]$ is both a surjective and a one-to-one correspondence function.

1.7   Inverse Function

Let $f : D \to E$. If for every y of E there exists a unique x of D such that $y = f(x)$, then $f(x)$ is called an **invertible function**, and $x = g(y) : E \to D$ is the **inverse function** of $f(x)$. That is, $f(g(y)) = y$ and $g(f(x)) = x$.

Examples:

a. The inverse function of $y = 2x + 3$ is $x = \dfrac{1}{2} y + \dfrac{3}{2}$

b. The inverse function of $y = x^2$, $x \geq 0$ is $x = \sqrt{y}$

   In the inverse function $x = g(y)$ the input variable is y and the output variable is x. It's more convenient to have x here as the input variable, so x and y are switched, and the result is the inverse function $y = g(x)$.

   In the previous example, the inverse function of $y = 2x + 3$ was $y = 0.5x + 1.5$.

Theorem: If $y = f(x)$ is invertible and $y = g(x)$ is its inverse function, then the graphs of these two functions are symmetric to each other about the line $y = x$ (Figure 8).

Theorem: The function $f : D \to E$ is invertible if, and only if, it is a one-to-one correspondence and a surjective function.

1.8   Equivalent Sets

a. Set A is called equivalent to set B if there exists a one-to-one correspondence and surjective function $f : A \to B$.

   Example: Interval $(-1,1)$ is equivalent to $\mathbb{R}$ as one-to-one correspondence function $\arctan \dfrac{\pi}{2} x$ mapping $(-1,1)$ on $\mathbb{R}$.

b. A set equivalent to natural numbers set $\mathbb{N}$ is called a **countable set**.

   Example: Sets $\mathbb{Z}$ and $\mathbb{Q}$ are countable sets.

c. A set equivalent to interval $(0,1)$ is called a **linear continuum**.

   Example: Open or closed intervals, $\mathbb{R}$, and $\mathbb{C}$ are continuum.

1.9   Operations with Functions

Let $f : D_1 \to R$, $g : D_2 \to R$.

Functions $f(x) + g(x)$, $f(x) \cdot g(x)$ and $\dfrac{f(x)}{g(x)}$ are only defined when x belongs to D_1 as well as to D_2, and therefore, are defined in domain $D = D_1 \cap D_2$.

To compose $f : D \to E$ and $g : E' \to F$, when $E \subset E'$, is function $h : D \to F$, which results from $h(x) = g(f(x)) = g \circ f(x)$.

Order of operations is significant. First, f is applied to x and, only then, g is applied to f.

Example: $f(x) = \sin x$, $g(x) = x^2 + 3x$, therefore

$g \circ f = (\sin x)^2 + 3\sin x$, $f \circ g = \sin(x^2 + 3x)$

1.10 Elementary Functions

The functions, resulting from the functions $f(x) = c$ (constant), $f(x) = x$, $f(x) = \sin x$, $f(x) = a^x$ by their addition, subtraction, multiplication, and partition, their compositions and their inverses, are **elementary functions**.

Examples:

a. Function $y = |x|$ is elementary since $y = \sqrt{x^2}$.

b. $\cos x = \sin\left(\dfrac{\pi}{2} - x\right)$ and   therefore, $y = \cos x$ is   an elementary function.

Chapter 2
Classes of Elementary Functions

2.1   Linear Function $y = ax + b$

2.2   Square Function $y = ax^2 + bx + c$

2.3   Power Function $y = x^\alpha$, $\alpha \in \mathbb{R}$

Examples are in Figures 1-4.

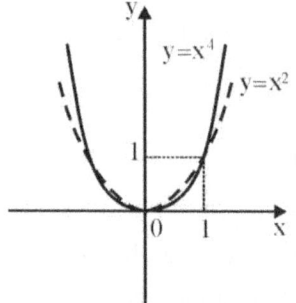

Figure 1

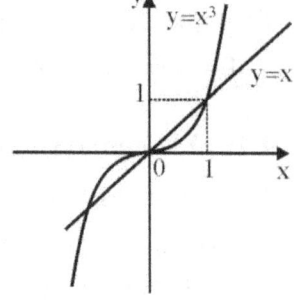

Figure 2

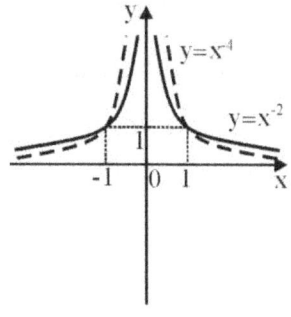

Figure 3

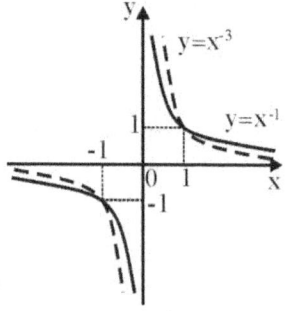

Figure 4

2.4   Exponential Function

$y = a^x$, $a > 0$, $a \neq 1$ (Figures 5, 6)

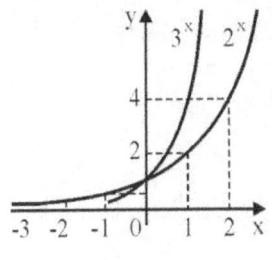

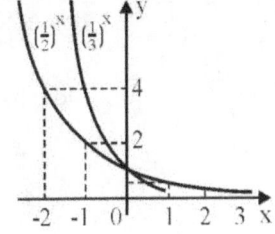

Figure 5   Figure 6

2.5   Logarithm Function

$y = \log_a x$, $a > 0$, $a \neq 1$. Graph is shown in Figure 7.

If its basis is $a = e$, it's denoted $y = \ln x$.

If its basis is $a = 10$, it's denoted $y = \lg x$

Function $y = \log_a x$ is the inverse of $y = a^x$. Their graphs are symmetric with respect to the straight line $y = x$ (Figure 8).

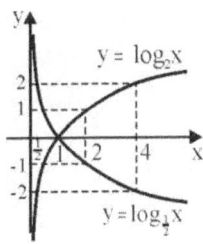

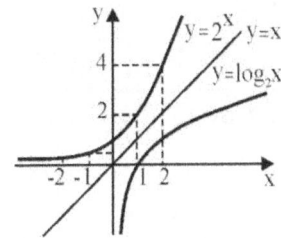

Figure 7   Figure 8

Basic formulas: $a > 0$, $a \neq 1$

$$\log_a (x \cdot y) = \log_a x + \log_a y \ , \ x > 0 \ , \ y > 0$$

$$\log_a \frac{x}{y} = \log_a x - \log_a y \ , \ x > 0 \ , \ y > 0$$

$$\log_a x^\alpha = \alpha \log_a x \ , \ x > 0 \ , \ \alpha \in \mathbb{R}$$

$$\log_a b = \frac{\log_c b}{\log_c a} \ , \ c > 0 \ , \ c \neq 1 \ \ ;$$

$$\log_a b = \frac{1}{\log_b a} \ , \ b > 0 \ , \ b \neq 1$$

2.6 Trigonometric Functions

a. Sine

$y = \sin x \ : \ \mathbb{R} \ \rightarrow \ [-1,1]$ is an odd periodic function with a 2π period (Figure 9).

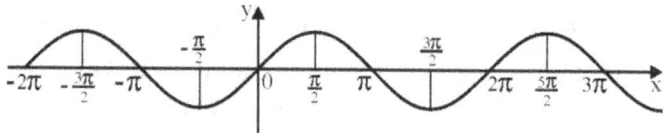

Figure 9

b. Cosine

$y = \cos x \ : \ \mathbb{R} \ \rightarrow \ [-1,1]$ is an even periodic function with a 2π period.

$\cos x = \sin (x + \frac{\pi}{2})$: Its graph results from shifting the graph $y = \sin x$ by $\frac{\pi}{2}$ leftward (Figure 10).

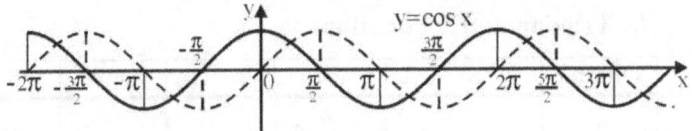

Figure 10

c. **Tangent**

$y = \tan x : \mathbb{R} - \left(\dfrac{\pi}{2} + \pi k \right) \to \mathbb{R}$, $k \in \mathbb{Z}$ is an odd periodic function with a π period (Figure 11).

d. **Cotangent**

$y = \cot x : \mathbb{R} - \pi k \to \mathbb{R}$, $k \in \mathbb{Z}$ is an odd periodic function with a π period (Figure 12).

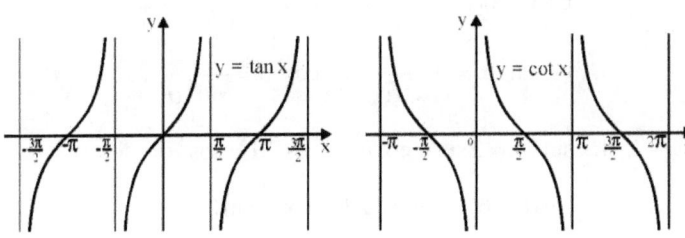

Figure 11   Figure 12

e. Increase and Decrease Domains and Denotations of Trigonometric Functions

Quarter	α	$\sin\alpha$	$\cos\alpha$	$\tan\alpha$	$\cot\alpha$
I	$0 < \alpha < \dfrac{\pi}{2}$	$0 \nearrow 1$	$1 \searrow 0$	$0 \nearrow \infty$	$\infty \searrow 0$
II	$\dfrac{\pi}{2} < \alpha < \pi$	$1 \searrow 0$	$0 \searrow -1$	$-\infty \nearrow 0$	$0 \searrow -\infty$
III	$\pi < \alpha < \dfrac{3}{2}\pi$	$0 \searrow -1$	$-1 \nearrow 0$	$0 \nearrow \infty$	$\infty \searrow 0$
IV	$\dfrac{3}{2}\pi < \alpha < 2\pi$	$-1 \nearrow 0$	$0 \nearrow 1$	$-\infty \nearrow 0$	$0 \searrow -\infty$

f. Basic Formulas

1. $\cos^2\alpha + \sin^2\alpha = 1$

2. $\tan\alpha \cdot \cot\alpha = 1$

3. $1 + \tan^2\alpha = \dfrac{1}{\cos^2\alpha}$, $1 + \cot^2\alpha = \dfrac{1}{\sin^2\alpha}$

4. $\sin 2\alpha = 2\sin\alpha\cos\alpha$, $\cos 2\alpha = \cos^2\alpha - \sin^2\alpha$

5. $\sin(\alpha \pm \beta) = \sin\alpha\cos\beta \pm \cos\alpha\sin\beta$

6. $\cos(\alpha \pm \beta) = \cos\alpha\cos\beta \mp \sin\alpha\sin\beta$

7. $\tan(\alpha \pm \beta) = \dfrac{\tan\alpha \pm \tan\beta}{1 \mp \tan\alpha \cdot \tan\beta}$, $\left(\alpha \pm \beta \neq \dfrac{\pi}{2} + \pi n\right)$

8. $2\cos\alpha \cdot \cos\beta = \cos(\alpha + \beta) + \cos(\alpha - \beta)$

9. $2\sin\alpha \cdot \cos\beta = \sin(\alpha + \beta) + \sin(\alpha - \beta)$

10. $2\sin\alpha \cdot \sin\beta = \cos(\alpha - \beta) - \cos(\alpha + \beta)$

11. $\cos\alpha + \cos\beta = 2\cos\dfrac{\alpha+\beta}{2}\cos\dfrac{\alpha-\beta}{2}$

12. $\cos\alpha - \cos\beta = -2\sin\dfrac{\alpha+\beta}{2}\sin\dfrac{\alpha-\beta}{2}$

13. $\sin\alpha \pm \sin\beta = 2\sin\dfrac{\alpha\pm\beta}{2}\cdot\cos\dfrac{\alpha\mp\beta}{2}$

14. $\tan\alpha \pm \tan\beta = \dfrac{\sin(\alpha\pm\beta)}{\cos\alpha\cos\beta}$

g. Reduction Formulas

	$-\alpha$	$90°\mp\alpha$ $\dfrac{\pi}{2}\mp\alpha$	$180°\mp\alpha$ $\pi\mp\alpha$	$270°\mp\alpha$ $\dfrac{3\pi}{2}\mp\alpha$	$360°\mp\alpha$ $2\pi\mp\alpha$
sin	$-\sin\alpha$	$\cos\alpha$	$\pm\sin\alpha$	$-\cos\alpha$	$\mp\sin\alpha$
cos	$\cos\alpha$	$\pm\sin\alpha$	$-\cos\alpha$	$\mp\sin\alpha$	$\cos\alpha$
tan	$-\tan\alpha$	$\pm\cot\alpha$	$\mp\tan\alpha$	$\pm\cot\alpha$	$\mp\tan\alpha$
cot	$-\cot\alpha$	$\pm\tan\alpha$	$\mp\cot\alpha$	$\pm\tan\alpha$	$\mp\cot\alpha$

h. Additional Formulas

1. $a\sin\alpha + b\cos\alpha = \sqrt{a^2+b^2}\,\sin(\alpha+\varphi)$, where φ is satisfy the equations $\sin\varphi = \dfrac{b}{\sqrt{a^2+b^2}}$, and $\cos\varphi = \dfrac{a}{\sqrt{a^2+b^2}}$

2. $\cos 3\alpha = 4\cos^3\alpha - 3\cos\alpha$, $\sin 3\alpha = 3\sin\alpha - 4\sin^3\alpha$

3. $1+\cos\alpha = 2\cos^2\dfrac{\alpha}{2}$, $1-\cos\alpha = 2\sin^2\dfrac{\alpha}{2}$

4. $\sin\alpha = \dfrac{2\tan\dfrac{\alpha}{2}}{1+\tan^2\dfrac{\alpha}{2}}$, $\cos\alpha = \dfrac{1-\tan^2\dfrac{\alpha}{2}}{1+\tan^2\dfrac{\alpha}{2}}$

$$5. \quad \tan\alpha = \frac{2\tan\frac{\alpha}{2}}{1-\tan^2\frac{\alpha}{2}} \ , \quad \cot\alpha = \frac{1-\tan^2\frac{\alpha}{2}}{2\tan\frac{\alpha}{2}}$$

i. Expressing Trigonometric Function Through Another

Computed through	$\sin\alpha$	$\cos\alpha$	$\tan\alpha$	$\cot\alpha$
$\sin\alpha =$	$\sin\alpha$	$\pm\sqrt{1-\cos^2\alpha}$	$\dfrac{\tan\alpha}{\pm\sqrt{1+\tan^2\alpha}}$	$\dfrac{1}{\pm\sqrt{1+\cot^2\alpha}}$
$\cos\alpha =$	$\pm\sqrt{1-\sin^2\alpha}$	$\cos\alpha$	$\dfrac{1}{\pm\sqrt{1+\tan^2\alpha}}$	$\dfrac{\cot\alpha}{\pm\sqrt{1+\cot^2\alpha}}$
$\tan\alpha =$	$\dfrac{\sin\alpha}{\pm\sqrt{1-\sin^2\alpha}}$	$\dfrac{\pm\sqrt{1-\cos^2\alpha}}{\cos\alpha}$	$\tan\alpha$	$\dfrac{1}{\cot\alpha}$
$\cot\alpha =$	$\dfrac{\pm\sqrt{1-\sin^2\alpha}}{\sin\alpha}$	$\dfrac{\cos\alpha}{\pm\sqrt{1-\cos^2\alpha}}$	$\dfrac{1}{\tan\alpha}$	$\cot\alpha$
$\sec\alpha =$	$\dfrac{1}{\pm\sqrt{1-\sin^2\alpha}}$	$\dfrac{1}{\cos\alpha}$	$\pm\sqrt{1+\tan^2\alpha}$	$\dfrac{\pm\sqrt{1+\cot^2\alpha}}{\cot\alpha}$
$\operatorname{cosec}\alpha$	$\dfrac{1}{\sin\alpha}$	$\dfrac{1}{\pm\sqrt{1-\cos^2\alpha}}$	$\dfrac{\pm\sqrt{1+\tan^2\alpha}}{\tan\alpha}$	$\pm\sqrt{1+\cot^2\alpha}$

In these formulas, the root's mark is the same as that of the left side of the function, which depends on the quarter being α .

Example: If $\dfrac{\pi}{2} < \alpha < \pi$, then when $\sin\alpha$ is computed through $\tan\alpha$, it is taken with $+$, since in quarter II function $\sin\alpha$ is positive, that is, $\sin\alpha = \dfrac{\tan\alpha}{1+\tan^2\alpha}$. Still, for the same angle, $\cos\alpha$ is taken with $-$, since, in the second quarter, $\cos\alpha < 0$, and therefore, $\cos\alpha = -\dfrac{1}{\sqrt{1+\tan^2\alpha}}$.

j. **Law of Sines**

For a triangle of a, b, c sides, and A, B, C angles, respectively

$$\frac{a}{\sin A} = \frac{b}{\sin B} = \frac{c}{\sin C} = 2R$$

where R is the radius of the circumscribed circle.

k. **Law of Cosines**

The square of one triangle side equals the sum of the squares of the other two sides minus twice the product of these two sides and the cosine of the angle between them:

$$a^2 = b^2 + c^2 - 2b \cdot c \cdot \cos A.$$

2.7   Inverse Trigonometric Functions

Trigonometric functions are periodical functions, and therefore not bijective. Reducing their natural domain, we get invertible functions:

a. The Inverse function of $y = \sin x$: $\left[-\dfrac{\pi}{2}, \dfrac{\pi}{2}\right] \rightarrow [-1,1]$ is

$y = \arcsin x \;:\; [-1,1] \;\rightarrow\; \left[-\dfrac{\pi}{2}, \dfrac{\pi}{2} \right]$ (arcsine x).

The graph of function $y = \arcsin x$ is symmetric to the graph of $y = \sin x$ with respect to the straight line $y = x$ (Figure 13).

$\arcsin x$ is an odd function.

b. The inverse of function $y = \cos x \;:\; [0, \pi] \rightarrow [-1,1]$ is $y = \arccos x \;:\; [-1,1] \rightarrow [0, \pi]$ (arccosine x) (Figure 14).

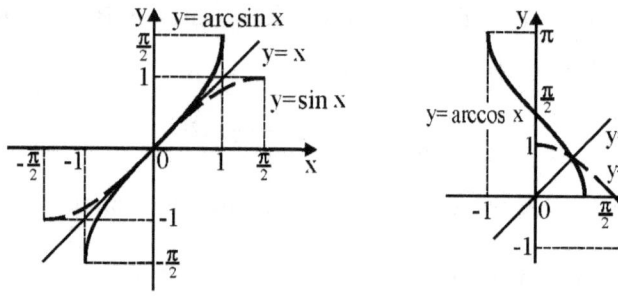

Figure 13   Figure 14

c. The inverse function of $y = \tan x \;:\; \left(-\dfrac{\pi}{2}, \dfrac{\pi}{2} \right) \rightarrow (-\infty, \infty)$ is

$y = \arctan x \;:\; (-\infty, \infty) \rightarrow \left(-\dfrac{\pi}{2}, \dfrac{\pi}{2} \right)$ (arctangent x).

$\arctan x$ is an odd function (Figure 15).

d. The inverse function of $y = \cot x \;:\; (0, \pi) \rightarrow (-\infty, \infty)$ is $y = \operatorname{arc cot} x \;:\; (-\infty, \infty) \rightarrow (0, \pi)$ (arccotangent x) (Figure 16).

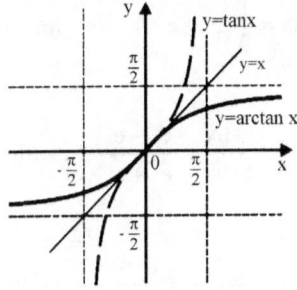

Figure 15   Figure 16

e. Additional formulas:

1. $\sin \arccos x = \sqrt{1 - x^2}$

2. $\cos \arcsin x = \sqrt{1 - x^2}$

3. $\sin \arctan x = \dfrac{x}{\sqrt{1 + x^2}}$

4. $\tan \arcsin x = \dfrac{x}{\sqrt{1 - x^2}}$

5. $\arcsin x + \arccos x = \dfrac{\pi}{2}$

6. $\arctan x + \operatorname{arc} \cot x = \dfrac{\pi}{2}$

2.8   Hyperbolic Functions

a. Hyperbolic sine: $\sin hx = \dfrac{1}{2}(e^x - e^{-x})$ is an odd function (Figure 17).

b. Hyperbolic cosine: $\cos hx = \dfrac{1}{2}(e^x + e^{-x})$ is an even function (Figure 17).

c. Hyperbolic tangent: $\tan hx = \dfrac{\sin hx}{\cos hx} = \dfrac{e^x - e^{-x}}{e^x + e^{-x}}$ is an odd function (Figure 18).

d. Hyperbolic cotangent: $\cot hx = \dfrac{1}{\tan hx} = \dfrac{e^x + e^{-x}}{e^x - e^{-x}}$ is an odd function (Figure 18).

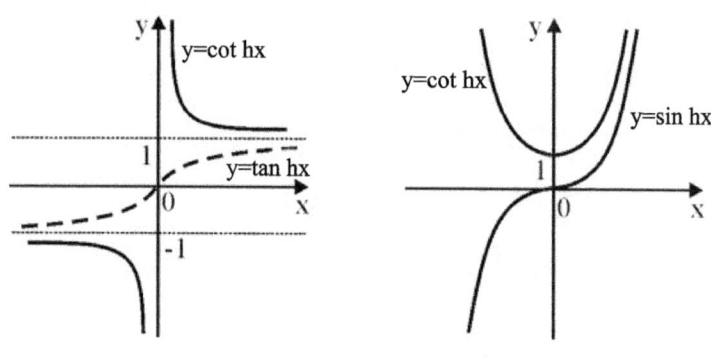

Figure 17   '   Figure 18

2.9   Basic Formulas

1. $\cos h^2 x - \sin h^2 x = 1$

2. $\tan hx \cdot \cot hx = 1$, $x \neq 0$

3. $\sin h(x \pm y) = \sin hx \cdot \cos hy \pm \cos hx \cdot \sin hy$

4. $\cos h(x \pm y) = \cos hx \cdot \cos hy \pm \sin hx \cdot \sin hy$

5. $\tan h(x \pm y) = \dfrac{\tan hx \pm \tan hy}{1 \pm \tan hx \cdot \tan hy}$

6. $\sin hx \pm \sin hy = 2\sin h\dfrac{x \pm y}{2} \cdot \cos h\dfrac{x \mp y}{2}$

7. $\cos hx + \cos hy = 2\cos h\dfrac{x+y}{2} \cdot \cos h\dfrac{x-y}{2}$

8. $\cos hx - \cos hy = 2\sin h\dfrac{x+y}{2} \cdot \sin h\dfrac{x-y}{2}$

9. $\cos h2x = \sin h^2 x + \cos h^2 x = 1 + 2\sin h^2 x = 2\cos h^2 x - 1$

10. $\sin h2x = 2\sin hx \cdot \cos hx$

11. $\tan h2x = \dfrac{2\tan hx}{1 + \tan h^2 x}$

Chapter 3
Parametric Form of a Function

In the explicit form of function $y = f(x)$, variables x and y relate directly.

But in its parametric representation, x and y relate indirectly. In this case, variables x and y are dependent of another variable, t, as presented by the two equations $x = \varphi(t)$, $y = \psi(t)$, in a given domain for variable t. Variable t is called the **parameter** of the form. If function $\varphi(t)$ is invertible above the given domain, then it can be denoted as $t = \varphi^{-1}(x)$, and be denoted in the explicit form of the function $y = \psi(t) = \psi(\varphi^{-1}(x))$.

Geometrical Interpretation

If parameter t is regarded as time variable, then equations $x = \varphi(t)$, $y = \psi(t)$ describe point (x, y) in a plane, where particle is situated at time t. Therefore, the plain curve consisting of all points $(\varphi(t), \psi(t))$ represents the trajectory of particle in a plane.

Examples:

a. $x = \cos t$, $y = \sin t$, $0 \le t \le \pi$ is a parametric representation of an upper semicircle, the explicit form of which is $y = \sqrt{1-x^2}$. If also the domain of t is extended to interval $[0, 2\pi]$, the result will be full circle $x^2 + y^2 = 1$, which is not a graph of a function.

b. $x = 3\cos t$. $y = 4\sin t$, $0 \le t \le \pi$ is a parametric form of a function. The function $\varphi(t) = 3\cos t$ is invertible in interval $[0, \pi]$. Then to directly relate between x and y, we extract t from the former equation. The result is $t = \arccos\dfrac{x}{3}$. Positioning it in the latter equation, we get

$$y = 4\sin\left(\arccos\frac{x}{3}\right) = 4\sqrt{1 - \frac{x^2}{9}}$$

Another way of finding it here is using the trigonometric identity

$$\sin^2 t + \cos^2 t = \left(\frac{x}{4}\right)^2 + \left(\frac{y}{3}\right)^2 = 1.$$

The result is $\dfrac{x^2}{16} + \dfrac{y^2}{9} = 1$, since $y = 3\cos t \ge 0$ for all t in its domain $[0, \pi]$, it is the upper half of the aforementioned ellipse $y = 4\sqrt{1 - \dfrac{x^2}{9}}$.

c. $x = t^3 - 1$, $y = t^2$ is a parametric form of function

$y = (x+1)^{\frac{2}{3}}$, since $x + 1 = t^3$, and therefore

$(x+1)^{\frac{2}{3}} = (t^3)^{\frac{2}{3}} = t^2 = y$.

Chapter 4
Functions in Polar Coordinate System

A function in a polar coordinate system (see III, 1.2) is a rule expressing the distance from the origin ρ as dependent of the variable of angle φ, $\rho = f(\varphi)$.

Notice: ρ must always be non-negative.

To draw the graph of that function, we can take any angle φ in the domain of the function, draw the appropriate ray and mark it with a point at a distance from the origin equal to $\rho = f(\varphi)$.

Examples:

a. $0 \le \varphi < 2\pi$, $\rho = 1$ for all angles φ. Since the distance from the origin is 1, it is a circle of radius 1 (Figure 1).

b. $\rho = \varphi$, $\varphi \ge 0$ **Archimedean Spiral** (Figure 2).

c. $\rho = 2\cos\varphi$, $-\dfrac{\pi}{2} \le \varphi < \dfrac{\pi}{2}$ a circle with radius 1, the center of which is at $(1,0)$ (Figure 3).

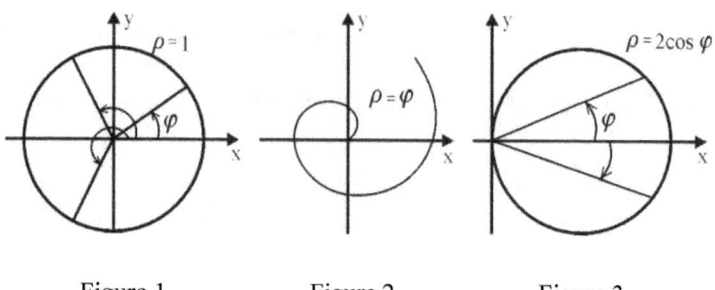

Figure 1   Figure 2   Figure 3

III. Analytic Geometry and Vectors

Chapter 1
Coordinate Systems in the Plane

1.1   Cartesian Coordinate System in a Plane

A **Cartesian coordinate system** consists of two number lines perpendicular to each other, where the horizontal is x and the vertical is y, intersecting at a point called the **origin**, and dividing the plane into four quarters. See Figure 1.

Each point M in the plane is specified by an ordered pair of numbers (a, b), where "a" stands for the distance of M from the y-axis, and marked with + (plus) if M is right of the y-axis, or with − (minus) if M s left of the y-axis. "b" is the distance of M from x-axis, and is marked with + (plus) if M is above the x-axis, or with − (minus) if M is below the x-axis. Thus, pair (a, b) is the **coordinates of point** M.

Figure 1 shows the coordinates of A(2,3), B(−1,2), C(3,−1), and D(−2,−2).

1.2   Polar Coordinate System

Let's fix point O in the plane, called a **pole**, and a ray $\overrightarrow{OP}$. The position of point M in the plane is strictly defined by its distance $\rho \geq 0$ from pole O and the positive angle φ between vector $\overrightarrow{OM}$ and axis $\overrightarrow{OP}$, when taken from $\overrightarrow{OP}$ to $\overrightarrow{OM}$, counterclockwise.

The pair ρ, φ are called the **polar coordinates** of point M.

All points in the plane have a unique pair of polar coordinates. Where $\rho \geq 0$ and $0 \leq \varphi < 2\pi$, Pole O is represented by $\rho = 0$ and any angle φ, which means the pole is not uniquely represented.

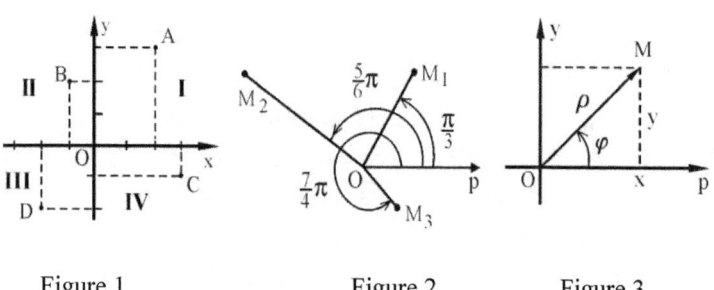

Figure 1   Figure 2   Figure 3

The polar coordinates of the points in Figure 2 are:

$$M_1 = \left(2, \frac{\pi}{3}\right) , \quad M_2 = \left(3, \frac{5\pi}{6}\right) , \quad M_3 = \left(1, \frac{7\pi}{4}\right)$$

1.3   Relation Between Cartesian and Polar Coordinates

Fixing Pole O at the origin of Cartesian system, and axis $\overrightarrow{OP}$ on its x-axis, we get the relations between the two coordinate systems (Figure 3).

$$x = \rho\cos\varphi \ , \ y = \rho\sin\varphi$$

And, vice versa, if x and y are known, the result should be

$$\rho = \sqrt{x^2 + y^2} \ , \ \sin\varphi = \frac{y}{\sqrt{x^2 + y^2}} \ , \ \cos\varphi = \frac{x}{\sqrt{x^2 + y^2}}$$

1.4   Distance Between Two Points

$$B(x_B, y_B) \ , \ A(x_A, y_A)$$

$$d = \sqrt{(x_A - x_B)^2 + (y_A - y_B)^2}$$

1.5   Area of a Triangle with Vertices

$$C(x_C, y_C) \ , \ B(x_B, y_B) \ , \ A(x_A, y_A)$$

$$S_{ABC} = \frac{1}{2}\,\text{mod}\begin{vmatrix} x_A - x_B & y_A - y_B \\ x_C - x_B & y_C - y_B \end{vmatrix} = \frac{1}{2}\,\text{mod}\begin{vmatrix} x_A & y_A & 1 \\ x_B & y_B & 1 \\ x_C & y_C & 1 \end{vmatrix} =$$

$$= \frac{1}{2}\,|(x_A - x_B)(y_C - y_B) - (y_A - y_B)(x_C - x_B)|$$

Chapter 2
Curves in the Plane

The equation $F(x,y) = 0$ in two variables x, y, is called a curve equation if, and only if, all points of the curve hold it. That is, points not on the curve do not hold this equation.
A major purpose of analytical geometry is analyzing curves through their equations.

2.1   Straight Line

First-order equation $ax + by + c = 0$ is a straight line equation. If φ is the angle between a straight line and the positive part of x-axis, measured from x-axis to the straight line counterclockwise, then the tangent of the angle is called the slope of the line.

The slope of a straight line is $k = \tan\varphi = -\dfrac{a}{b}$, $(b \neq 0)$.

If $b = 0$, the line has no slope.

a. $y = kx + b$ is the equation of a line with a slope k.

b. $y = b$ is a line parallel to the x-axis (its slope is $k = 0$)

c. $x = a$ is a line parallel to y-axis (it has no slope).

d. $y = kx$ is a straight line passing through the origin.

e.   The equation of a line passing through points (x_1, y_1) and (x_2, y_2) is

$$y - y_1 = \frac{y_2 - y_1}{x_2 - x_1} \cdot (x - x_1).$$

f. The distance of point (x_0, y_0) from the straight line $ax + by + c = 0$ is

$$d = \frac{|ax_0 + by_0 + c|}{\sqrt{a^2 + b^2}}.$$

g. Let there be two straight lines, $y = k_1 x + b_1$, and $y = k_2 x + b_2$.

1. If $k_1 = k_2$, then they are parallel to each other.

2. If $k_1 \cdot k_2 = -1$, then they are perpendicular to each other.

3. $\tan \alpha = \frac{k_2 - k_1}{1 + k_1 k_2}$, where α is the angle between the lines.

2.2   Circles

a. A **circle** is the locus of points for which the distance from a given point is constant, called a **radius**.

b. The equation of circle for which the center is in point (a, b) and its radius is R is

$$(*) \quad (x - a)^2 + (y - b)^2 = R^2$$

c. $x^2 + y^2 + ax + by + c = 0$ is an equation of a circle if $a^2 + b^2 - 4c > 0$. In such a case, its center is point $\left(-\frac{a}{2}, -\frac{b}{2}\right)$, and its radius is

$$\frac{1}{2}\sqrt{a^2 + b^2 - 4c}$$

d. The equation of a tangent line to circle (*) at the point (x_0, y_0) is

$$(x-a)(x_0-a)+(y-b)(y_0-b)=R^2$$

e. Straight line $y=kx+n$ is tangent to circle (*) if, and only if

$$(b-ak-n)^2 = R^2(1+k^2)$$

2.3   Canonical Ellipses

a. An **ellipse** is the locus of points M for which the sum of distance to two points F_1 and F_2, called **foci**, is constant $F_1M+F_2M=2a$.

b. An ellipse is canonical if its foci are on x-axis and symmetric to each other about y-axis.

c. Canonical ellipse equation: $\dfrac{x^2}{a^2}+\dfrac{y^2}{b^2}=1$, where $F_1(c,0)$, $F_2(-c,0)$ are its foci, a, and b are its semi-axes, and $b^2 = a^2 - c^2$.

d. The equation of a tangent line to ellipse at the point $M(x_0,y_0)$ is $\dfrac{xx_0}{a^2}+\dfrac{yy_0}{b^2}=1$.

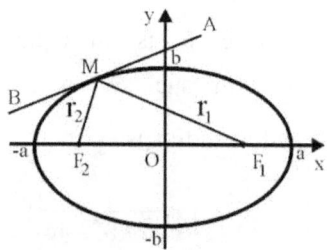

e. Its eccentricity is $\varepsilon = \dfrac{c}{a} = \dfrac{\sqrt{a^2 - b^2}}{a} < 1$.

f. $r_1 = F_1 M$, $r_2 = F_2 M$ are focal radii for point $M(x_M, y_M)$, such that $r_1 = a - \varepsilon x_M$, $r_2 = a + \varepsilon x_M$.

g. The tangent to ellipse forms equal angles with the focal radii of point of tangency, such that $\angle AMF_1 = \angle BMF_2$.

h. The straight line $y = kx + n$ is tangent to canonical ellipse if, and only if, $n^2 = k^2 a^2 + b^2$.

i. $\rho = \dfrac{b^2}{a(1 - \varepsilon \cos \varphi)}$ is the equation of ellipse in polar coordinate system.

2.4   Hyperbolas

a. A **hyperbola** is a locus of points for each of which the absolute value of the difference between the distances to two given points F_1 and F_2, called foci, is constant $|F_1 M - F_2 M| = 2a$.

b. A hyperbola is canonical if its foci are on the x-axis and symmetric to each other about y-axis.

c. A canonical hyperbola equaton is $\dfrac{x^2}{a^2} - \dfrac{y^2}{b^2} = 1$, where the foci are $F_1(c, 0)$, $F_2(-c, 0)$ (Figure 1).

   a is its real semi-axis, while b is its imaginary semi-axis, and $b^2 = c^2 - a^2$.

d. The lines $y = -\dfrac{b}{a} x$ and $y = \dfrac{b}{a} x$ are called the asymptotes of the hyperbola.

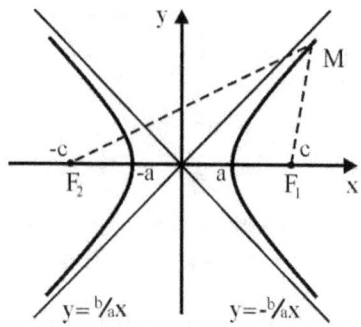

Figure 1

e. $\dfrac{x^2}{a^2} - \dfrac{y^2}{b^2} = -1$ is the conjugate hyperbola of a canonical hyperbola (Figure 2).

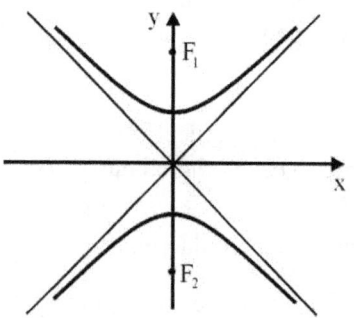

Figure 2

f. Its eccentricity is $\varepsilon = \dfrac{c}{a} = \dfrac{\sqrt{a^2 + b^2}}{a} > 1$.

g. $r_1 = F_1M$, $r_2 = F_2M$ are the focal radii of point $M(x_M, y_M)$, such that $r_1 = \varepsilon x + a$, $r_2 = a - \varepsilon x$.

h. The tangent to a hyperbola at point of tangency M bisects the angle $\angle F_1 M F_2$.

i. The equation of tangent line to canonical hyperbola at the point (x_0, y_0) is $\dfrac{xx_0}{a^2} - \dfrac{yy_0}{b^2} = 1$.

j. Line $y = kx + n$ is tangent to canonical hyperbola if, and only if, $n^2 = k^2 a^2 - b^2$.

k. $\rho = \dfrac{b^2}{a(1 - \varepsilon \cos \varphi)}$ is the canonical parabola equation relative to a polar coordinate system.

2.5   Parabolas

a. A **parabola** is the locus of points for which the distance to a given point F, called parabolic focus is equal to their distance to a given straight line called **directrix**, such that $MA = MF$.

b. A parabola is called canonical if its focus F is on the x-axis, its directrix is perpendicular to the x-axis, and the focus and perpendicular are on both sides of the origin, at equal distances from it.

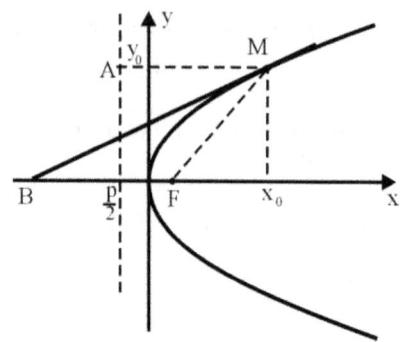

c. The canonical parabola equation is $y^2 = 2px$, where the focus is at $F\left(\dfrac{p}{2}, 0\right)$ and the equation of its directrix is $x = -\dfrac{p}{2}$.

d. The equation of a tangent line to a parabola at point of tangency $M(x_0, y_0)$ is $y_0 = p(x + x_0)$.

e. Straight line $y = kx + n$ is a tangent to canonical parabola if, and only if, $p = 2kn$.

f. $FM = r$ is the focal radius of $M(x_0, y_0)$, $r = x_0 + \dfrac{p}{2}$.

g. $\angle FMB = \angle FBM$ if MB is tangent to the parabola.

h. $\rho = \dfrac{p}{1 - \cos\varphi}$ is the equation of parabola relative to polar coordinates.

2.6   Conic Sections

Circle, ellipse, hyperbola, parabola, and two intersecting straight lines are obtained by intersection of a plane with a right circular cone.

Theorem: Let β be the angle between the generator of the cone and its axis. Then:

a. If a plane does not pass through the vertex of a cone, forming angle α with the cone axis, then the line of intersection is

   1. A circle, when $\alpha = \frac{1}{2}\pi$

   2. An ellipse, when $0 < \beta < \alpha < \frac{1}{2}\pi$ (Figure 1).

   3. A parabola, when $\alpha = \beta$ (Figure 2).

   4. A hyperbola, when $0 \le \alpha < \beta$ (Figure 3).

   5. Two intersecting straight lines, when $\alpha = 0$ (Figure 4).

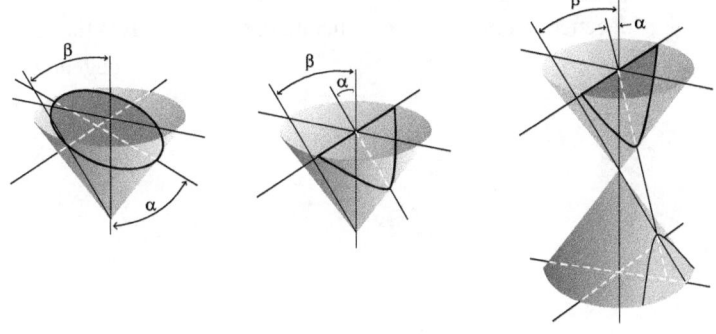

Figure 1   Figure 2   Figure 3

b. If the plane passes through the vertex, a point, or a straight line, a pair of straight lines may be obtained by the intersection (Figure 4).

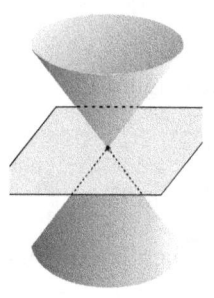
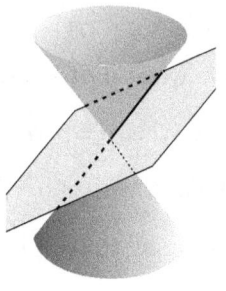
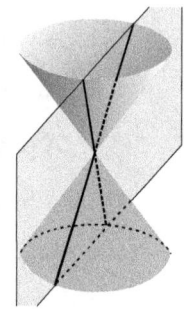

A point, or a degenerate circle

A straight line, or a degenerate parabola

Intersection of two lines, or a degenerate hyperbola

Figure 4

Chapter 3
Second-Order Curves in Plane - General Theory

(*) $ax^2 + 2bxy + cy^2 + 2dx + 2ey + f = 0$

is a general equation of second-order curve with **two variables**.

By switching the variables, that is, rotating the axis, we get

$$\begin{cases} x = X\cos\alpha - Y\sin\alpha \\ y = X\sin\alpha + Y\cos\alpha \end{cases}$$

When α is the solution of the equation $b\cos 2\alpha - \tfrac{1}{2}(a-c)\sin 2\alpha = 0$, we can turn (*) into an equation where the coefficient of xy equals to zero

$$a'X^2 + c'Y^2 + 2d'X + 2e'Y + f' = 0$$

It is denoted the following way

$$a'\left(X + \frac{d'}{a'}\right)^2 + c'\left(Y + \frac{e'}{c'}\right)^2 + f' - \left(\frac{d'}{a'}\right)^2 - \left(\frac{e'}{c'}\right)^2 = 0$$

After shifting the origin to point $\left(-\dfrac{d'}{a'}, -\dfrac{e'}{c'}\right)$, its canonical form is obtained $a'X'^2 + c'Y'^2 + f'' = 0$

Theorem: Denote

$$I_1 = a + c \quad I_2 = \begin{vmatrix} a & b \\ b & c \end{vmatrix} \quad I_3 = \begin{vmatrix} a & b & d \\ b & c & e \\ d & e & f \end{vmatrix} \quad I_4 = \begin{vmatrix} a & d \\ d & f \end{vmatrix} + \begin{vmatrix} c & e \\ e & f \end{vmatrix}$$

Let λ_1 and λ_2 be the solutions of equation

$$\lambda^2 - (a+c)\lambda + (ac - b^2) = 0$$

(That is, λ_2, λ_1 are the eigenvalues of I_2). Therefore:

1. If $I_2 \neq 0$, after rotating and shifting the axis, we obtain from equation (*) the following:

$$\lambda_1 X^2 + \lambda_2 Y^2 + \frac{I_3}{I_2} = 0$$

   a. If $I_2 > 0$ and $I_1 \cdot I_3 < 0$, then the curve is an ellipse.

   b. If $I_2 > 0$ and $I_1 \cdot I_3 > 0$, then there are no such points. The intersection is an imaginary ellipse.

   c. If $I_2 > 0$ and $I_3 = 0$, then it is a point.

   d. If $I_2 < 0$ and $I_3 \neq 0$, then the curve is a hyperbola.

   e. If $I_2 < 0$ and $I_3 = 0$, then it is two intersecting straight lines.

2. If $I_2 = 0$ and $I_3 \neq 0$, and the equation obtained from (*) is

   $I_1 Y^2 + 2\sqrt{-\dfrac{I_1}{I_3}} X = 0$, the curve is a parabola.

3. If $I_3 = I_2 = 0$, then the equation obtained is $I_1^2 Y^2 + I_4 = 0$

   a. If $I_4 < 0$, then it forms two parallel straight lines.

   b. If $I_4 = 0$, then it forms two coincident lines.

   c. If $I_4 > 0$, then the equation holds for no point.

Chapter 4
Coordinate Systems in Space, Space Vectors

4.1   Vector Concept

a. A **vector** is a defined quantity with both direction and magnitude.

b. It is graphically represented by a directional segment, that is, a segment with an arrow on one end, representing the direction of the vector.

c. The absolute value of a vector is called the length or a vector.

d. Vectors are denoted by small, bold Latin characters $\mathbf{c}, \mathbf{b}, \mathbf{a}$, and modules are denoted as $|\mathbf{c}|, |\mathbf{b}|, |\mathbf{a}|$. A vector starting at point A and ending at point B is denoted as $\overrightarrow{AB}$ and its length is denoted as AB or $|\overrightarrow{AB}|$.

e. $\overrightarrow{BA}$ is the opposite vector of $\overrightarrow{AB}$ so they are denoted as $\overrightarrow{AB} = -\overrightarrow{BA}$.

f. A vector of length 1 is called unit vector.

g. Two vectors are equal if their lengths are equal and their directions are identical.

h. The angle between vectors is the angle between their positive-direction parts. So, if vectors $\mathbf{a}$ and $\mathbf{b}$ are parallel to each other, that is, they are situated on parallel lines, and have the same direction, the angle between them is zero. If vectors $\mathbf{a}$ and $\mathbf{b}$ are parallel to each other and are of opposite directions, the angle between them is $180°$.

4.2   Vector Algebra

a. Vector additions: vector **c**, the sum of vectors **a** and **b**, is denoted **c=a+b**, and defined the following way: displace vectors **a** and **b** to common initial point A. Vector **c** is the diagonal $\overrightarrow{AB}$ of parallelogram constructed in vectors **a** and **b** from A to B.

Vector **c** can also be constructed this way: Draw **a**; from the end of **a**, draw **b**. Connecting the initial point of **a** to the end of **b**, we obtain vector **a+b**.

b. Adding more than two vectors is defined similarly.

c. If α is the angle between vectors **a** and **b**, then:

$$|a+b|=\sqrt{|a|^2+|b|^2-2|a|\cdot|b|\cdot\cos\alpha}\,.$$

d. Vector subtraction: The difference of vectors **a** and **b** is vector $d=a-b$, for which holds $d+b=a$.

e. The projection of vector **a** on **m** is scalar a_m which equals to the length of directed segment $\overrightarrow{A'B'}$, when A' is the projection of the initial point and B', the projection of the terminal point of **a**, marked with plus when $\overrightarrow{A'B'}$ is at the direction of **m**, and with minus when it is in the opposite direction of **m**.

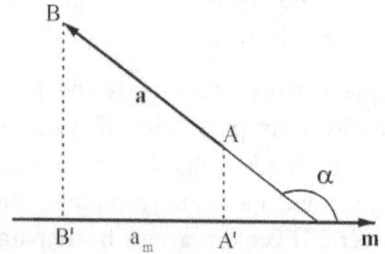

f. The projection of the sum of two vectors equals to the sum of their projections.

g. The projection of a difference of two vectors equals to the difference of their projections.

4.3   Vector Triangle Inequality

For all two vectors **a** and **b** there holds

$$|a| + |b| \geq |a + b| \geq \big||a| - |b|\big|$$

4.4   Cartesian Coordinate System in Space

a. Unit vectors **k,j,i** define a right-handed Cartesian system, if:

1. They are perpendicular to each other

$$k \perp i \; , \; j \perp k \; , \; i \perp j$$

2. All three vectors are initiated in one point, and, the motion from **i** to **j** counterclockwise is visible from the terminal point of vector **k** .

b. **Cartesian coordinate system in space** consists of three axes originating from one point O: x-axis at the direction of vector **i**; y-axis at the direction of vector **j**, and z-axis at the direction of vector **k** .

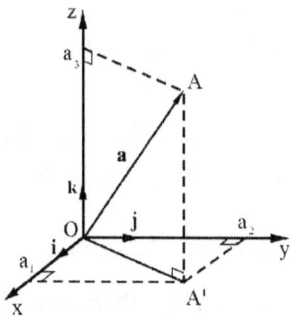

c. Every vector **a** can be represented in the form of $\mathbf{a} = a_1\mathbf{i} + a_2\mathbf{j} + a_3\mathbf{k}$. Three entities (a_1, a_2, a_3) are called **Cartesian coordinates**.

d. The space coordinates of point A are the space coordinates of vector $\overrightarrow{OA}$.

e. For every point in space there is a corresponding ordered set of three real numbers, and for every ordered set of three real numbers, there is a corresponding point in space.

f. **Example**: In the following Figure, there are points $D(3,3,-2), C(1,6,0), B(2,-5,3), A(-2,5,4)$

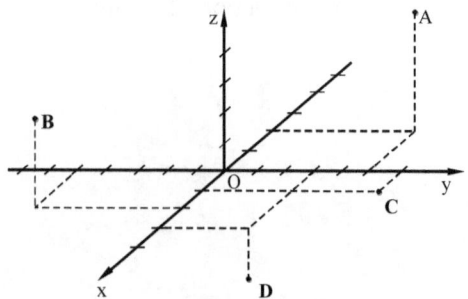

4.5   Vector in Coordinate System

Let's give vectors $\mathbf{a} = (a_1, a_2, a_3)$, $\mathbf{b} = (b_1, b_2, b_3)$, and scalar α, then:

a. $\alpha \mathbf{a} = (\alpha a_1, \alpha a_2, \alpha a_3)$

b. $|\mathbf{a}| = \sqrt{a_1^2 + a_2^2 + a_3^2}$

c. $\mathbf{a} + \mathbf{b} = (a_1 + b_1, a_2 + b_2, a_3 + b_3)$

d. $\mathbf{a} - \mathbf{b} = (a_1 - b_1, a_2 - b_2, a_3 - b_3)$

e. $\mathbf{a}$ is parallel to $\mathbf{b}$ $(\mathbf{a} \,\|\, \mathbf{b})$ if, and only if, $\dfrac{a_1}{b_1} = \dfrac{a_2}{b_2} = \dfrac{a_3}{b_3}$

4.6   Vector Direction in Space

a.   **Direction angles** of vector $\mathbf{a}$, are angles α, β, γ it forms with the positive direction of axes x, y, z, respectively.

b.   **Direction cosines** of $\mathbf{a}$ are $\cos\alpha, \cos\beta, \cos\gamma$.

c.   If $\mathbf{a} = (a_1, a_2, a_3)$, then

$$\cos\alpha = \frac{a_1}{|\mathbf{a}|} \;,\quad \cos\beta = \frac{a_2}{|\mathbf{a}|} \;,\quad \cos\gamma = \frac{a_3}{|\mathbf{a}|}.$$

d. $\cos^2\alpha + \cos^2\beta + \cos^2\gamma = 1$

4.7   Inner (Scalar) Product

a.   The inner product of two vectors $\mathbf{a}$, $\mathbf{b}$ is scalar $\mathbf{a} \cdot \mathbf{b}$ ($\mathbf{a}$ dot $\mathbf{b}$), defined as:

$$\mathbf{a} \cdot \mathbf{b} = |\mathbf{a}| \cdot |\mathbf{b}| \cdot \cos\alpha$$

when α is the angle between the vectors.

b. $\mathbf{a} \cdot \mathbf{b} = |\mathbf{a}| \cdot b_a = |\mathbf{b}| \cdot a_b$, when a_b is the projection of $\mathbf{a}$ on $\mathbf{b}$, and b_a is the projection of $\mathbf{b}$ on $\mathbf{a}$.

c. Vectors **a** and **b** are mutually perpendicular if, and only if, $\mathbf{a} \cdot \mathbf{b} = 0$.

d. If $\mathbf{a} = (a_1, a_2, a_3)$ and $\mathbf{b} = (b_1, b_2, b_3)$, then:

$$\mathbf{a} \cdot \mathbf{b} = a_1 b_1 + a_2 b_2 + a_3 b_3 .$$

e. $\mathbf{a} \cdot \mathbf{b} = \mathbf{b} \cdot \mathbf{a}$, $\alpha(\mathbf{a} \cdot \mathbf{b}) = (\alpha \mathbf{a}) \cdot \mathbf{b} = \mathbf{a} \cdot (\alpha \mathbf{b})$, (α is a scalar).

f. $\mathbf{a} \cdot (\mathbf{b} + \mathbf{c}) = \mathbf{a} \cdot \mathbf{b} + \mathbf{a} \cdot \mathbf{c}$.

g. $\mathbf{a} \cdot \mathbf{a} = \mathbf{a}^2 = |\mathbf{a}|^2$.

4.8   Cross (Vector) Product

a. Cross product of vectors **a**, **b** is vector **c** denoted as $\mathbf{a} \times \mathbf{b}$ (**a** cross **b**), which holds:

1. $|\mathbf{c}| = |\mathbf{a} \times \mathbf{b}| = |\mathbf{a}| \cdot |\mathbf{b}| \cdot \sin\alpha$, when α is the angle between vectors **a**, **b**. In other words, the length of vector $\mathbf{c} = \mathbf{a} \times \mathbf{b}$ equals the area of the parallelogram with vectors for sides.

2. Vector **c** is perpendicular to the plane of vectors **a** and **b**, and its direction is such that, from the terminal point of **c** the motion from **a** to **b** counterclockwise is visible (right-hand rule).

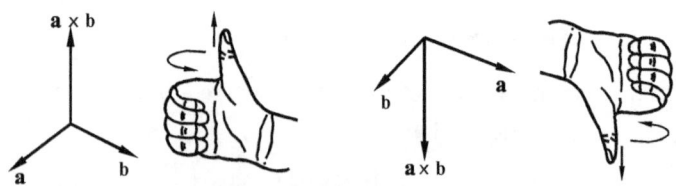

b. If $\mathbf{a} = (a_1, a_2, a_3)$ and $\mathbf{a} = (b_1, b_2, b_3)$, then

$$\mathbf{a} \times \mathbf{b} = \begin{vmatrix} \mathbf{i} & \mathbf{j} & \mathbf{j} \\ a_1 & a_2 & a_3 \\ b_1 & b_2 & b_3 \end{vmatrix}$$

$$= (a_2 b_3 - a_3 b_2) \mathbf{i} + (a_3 b_1 - a_1 b_3) \mathbf{j} + (a_1 b_2 - a_2 b_1) \mathbf{k}$$

c. Vectors $\mathbf{a}$ and $\mathbf{b}$ are collinear if, and only if, $\mathbf{a} \times \mathbf{b} = \mathbf{0}$.

d. $\mathbf{a} \times \mathbf{b} = -\mathbf{b} \times \mathbf{a}$.

e. $(\mathbf{a} + \mathbf{b}) \times \mathbf{c} = \mathbf{a} \times \mathbf{c} + \mathbf{b} \times \mathbf{c}$.

f. $\alpha(\mathbf{a} \times \mathbf{b}) = (\alpha \mathbf{a}) \times \mathbf{b} = \mathbf{a} \times (\alpha \mathbf{b})$.

4.9   Vector Triple Product

a. The multiplication of $\mathbf{a} \cdot (\mathbf{b} \times \mathbf{c})$, of vectors $\mathbf{a} = (a_1, a_2, a_3)$, $\mathbf{b} = (b_1, b_2, b_3)$, $\mathbf{c} = (c_1, c_2, c_3)$ is called triple product.

b. $\mathbf{a} \cdot (\mathbf{b} \times \mathbf{c}) = \begin{vmatrix} a_1 & a_2 & a_3 \\ b_1 & b_2 & b_3 \\ c_1 & c_2 & c_3 \end{vmatrix}$

c. $\mathbf{a} \cdot (\mathbf{b} \times \mathbf{c}) = (\mathbf{a} \times \mathbf{b}) \cdot \mathbf{c}$

d. $\mathbf{a} \cdot (\mathbf{b} \times \mathbf{c})$ is a scalar modulus of which equals to the volume of parallelepiped constructed on vectors $\mathbf{a}, \mathbf{b}, \mathbf{c}$.

e. Vectors $\mathbf{a}, \mathbf{b}, \mathbf{c}$ are coplanar, or situated on the same plane, if, and only if, $\mathbf{a} \cdot (\mathbf{b} \times \mathbf{c}) = 0$.

Chapter 5
Plane in Space

5.1   Plane

The equation of plane passing, through point $M_0(x_0, y_0, z_0)$ and perpendicular to vector $N = (A, B, C)$ is

$$A(x - x_0) + B(y - y_0) + c(z - z_0) = 0$$

Vector N is a vector **normal** to the plane.

5.2   General Form of a Plane Equation

$$Ax + By + Cz + D = 0$$

5.3   Distance of a Point to the Plane

The distance of point $M_0(x_0, y_0, z_0)$ to plane $Ax + By + Cz + D = 0$ equals

$$d = \frac{|Ax_0 + By_0 + Cz_0 + D|}{\sqrt{A^2 + B^2 + C^2}}.$$

5.4   Parallel Planes

a. Planes $A_1 x + B_1 y + C_1 z + D_1 = 0$ and
$A_2 x + B_2 y + C_2 z + D_2 = 0$ are parallel to each other if, and only if, their normal vectors $N_1 = (A_1, B_1, C_1)$ and $N_2 = (A_2, B_2, C_2)$ are parallel to each other.

b. These planes are parallel to each other if, and only if,

$$\frac{A_1}{A_2} = \frac{B_1}{B_2} = \frac{C_1}{C_2}.$$

5.5   Angle Between Two Planes

The angle of two non-parallel planes equals to the angle of their normal vectors N_1 and N_2 or to the supplementary angle of it. Therefore, the cosine of the acute angle between the planes is

$$\cos \alpha = \frac{|N_1 \cdot N_2|}{|N_1| \cdot |N_2|}$$

Or, in its coordinate form

$$\cos \alpha = \frac{|A_1 \cdot A_2 + B_1 \cdot B_2 + C_1 \cdot C_2|}{\sqrt{A_1^2 + B_1^2 + C_1^2} \cdot \sqrt{A_2^2 + B_2^2 + C_2^2}}$$

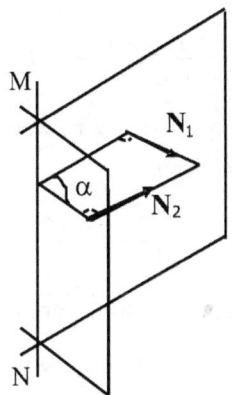

Chapter 6
Straight Line in Space

6.1   Straight Line Equation

The equation of a straight line passing through point $M_0(x_0, y_0, z_0)$ and is parallel to given vector $\mathbf{a} = (k, m, n)$ is

a. In its **parametric form**

$$\begin{cases} x = x_0 + tk \\ y = y_0 + tm \\ z = z_0 + tn \end{cases} , \quad -\infty < t < \infty$$

when t is **parameter**.

b. In its **symmetric** or **canonical form**

$$\frac{x - x_0}{k} = \frac{y - y_0}{m} = \frac{z - z_0}{n} \quad , \quad (k \neq 0 \, , \, n \neq 0 \, , \, m \neq 0)$$

If one of the numbers n, m, k is zero, for instance $m = 0$, then the equation is

$$\frac{x - x_0}{k} = \frac{z - z_0}{n} \quad , \quad y = y_0$$

Similarly, canonical forms are obtained if $k = 0$ or $n = 0$.

c. In its **vector form**: $\overrightarrow{OM} = \overrightarrow{OM_0} + t\mathbf{a}$ or

$$(x, y, z) = (x_0, y_0, z_0) + t(k, m, n), \quad -\infty < t < \infty .$$

6.2   Straight Line as Intersection of Two Planes

Two non-parallel planes intersect in a straight line, and therefore, system of equations

$$(*) \begin{cases} A_1 x + B_1 y + C_1 z + D_1 = 0 \\ A_2 x + B_2 y + C_2 z + D_2 = 0 \end{cases}$$ is a straight line equation.

Now we shall see how the canonical equation of the straight line can be obtained from (*):

$N_1 = (A_1, B_1, C_1)$ and $N_2 = (A_2, B_2, C_2)$ are vectors normal to the given planes respectively. The intersection line of the two planes is on each of them, and is therefore perpendicular to N_1, N_2. Therefore, the direction vector $\mathbf{a}$ of the straight line is $\mathbf{a} = N_1 \times N_2$.

A certain solution of the system describes a point on the intersection line of the two planes. That is, we found vector $\mathbf{a}$ and a point on the strait line, and therefore, the straight line equation can be denoted either in its canonical or parametric form.

Example: Let's get the canonical equation of the given line through planes $x - y + 3z - 3 = 0$, $3x + 2y - z - 2 = 0$.

The normal vectors to these planes are
$N_2(1, -1, 3)$, $N_1 = (3, 2, -1)$.
Let us calculate $\mathbf{a} = N_1 \times N_2 = 5\mathbf{i} - 10\mathbf{j} - 5\mathbf{k}$.
To find a point on the line, we substitute $z = 0$ in the planes equation, the following is resulted from the planes equations

$$\begin{cases} x - y - 3 = 0 \\ 3x + 2y - 2 = 0 \end{cases} \rightarrow x = 1.6 \ , \ y = -1.4 .$$

Point $M_0(1.6, -1.4, 0)$ is on the intersection line of the planes, so $\dfrac{x - 1.6}{5} = \dfrac{y + 1.4}{-10} = \dfrac{z}{-5}$ is the intersection line equation.

6.3   Straight Line Passing Through Two Points

$M_0(x_0, y_0, z_0)$ and $M_1(x_1, y_1, z_1)$:

Vector $\overrightarrow{M_0 M_1} = (x_1 - x_0, y_1 - y_0, z_1 - z_0)$ is in the direction of the straight line, so its equation is

$$\frac{x - x_0}{x_1 - x_0} = \frac{y - y_0}{y_1 - y_0} = \frac{z - z_0}{z_1 - z_0}$$

or, in its vector form

$$(x, y, z) = (x_0, y_0, z_0) + t(x_1 - x_0, y_1 - y_0, z_1 - z_0), \quad -\infty < t < \infty$$

6.4   Distance of a Point From a Straight Line

The distance of point M from a straight line passing through point M_0 and parallel to vector $\mathbf{a}$ is

$$d = \frac{|\overrightarrow{M_0 M} \times \mathbf{a}|}{|\mathbf{a}|}.$$

6.5   Angle Between Two Straight Lines

$$\frac{x - x_0}{k} = \frac{y - y_0}{m} = \frac{z - z_0}{n} \quad , \quad \frac{x - x_1}{k_1} = \frac{y - y_1}{m_1} = \frac{z - z_1}{n_1}$$

The angle between the lines is the angle between vectors $\mathbf{a} = (k, m, n)$ and $\mathbf{a}_1 = (k_1, m_1, n_1)$.

$$\cos\alpha = \frac{\mathbf{a} \cdot \mathbf{a}_1}{|\mathbf{a}| \cdot |\mathbf{a}_1|} = \frac{kk_1 + mm_1 + nn_1}{\sqrt{k^2 + m^2 + n^2} \cdot \sqrt{k_1^2 + m_1^2 + n_1^2}}$$

Two straight lines are perpendicular to each other if, and only if, $\cos\alpha = 0$, or

$$kk_1 + mm_1 + nn_1 = 0.$$

Two straight lines are parallel to each other if, and only if,

$$\frac{k}{k_1} = \frac{m}{m_1} = \frac{n}{n_1}$$

Chapter 7
Straight Line and Plane

7.1   Mutual Position of Straight Line and Plane

Let there be straight line such that $\dfrac{x-x_0}{k} = \dfrac{y-y_0}{m} = \dfrac{z-z_0}{n}$ and plane such that $Ax + By + Cz + D = 0$. Let $\mathbf{a} = (k, m, n)$ be the direction vector of the line, and $\mathbf{N} = (A, B, C)$ the normal to the plane. Therefore:

a. The straight line **is parallel to the plane** if, and only if, $\mathbf{N} \perp \mathbf{a}$, or

$$\mathbf{N} \cdot \mathbf{a} = Ak + Bm + Cn = 0$$

b. The straight line **is perpendicular to the plane** if, and only if, $\mathbf{N} \parallel \mathbf{a}$, or $\dfrac{A}{k} = \dfrac{B}{m} = \dfrac{C}{n}$

c. The angle between the straight line the plane is angle α between the straight line and its projection on the plane (see next Figure).

$$\sin \alpha = \cos \beta = \frac{\mathbf{a} \cdot \mathbf{N}}{|\mathbf{a}| \cdot |\mathbf{N}|} = \frac{Ak + Bm + Cn}{\sqrt{A^2 + B^2 + C^2} \cdot \sqrt{k^2 + m^2 + n^2}}$$

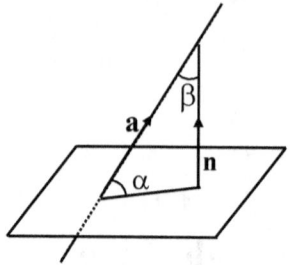

7.2   Intersection Point of a Straight Line and a Plane

To find the intersection point, we should write the straight line equation in its parametric form

$$(*) \quad z = z_0 + nt \ , \ y = y_0 + mt \ , \ x = x_0 + kt.$$

By substitution in the plane equation, we get an equation related to t

$$(Ak + Bm + Cn)t + Ax_0 + By_0 + Cz_0 + D = 0.$$

By extracting t, denoting it with (*), we obtain the intersection point.

Chapter 8
Canonical Forms of Second-Order Surfaces in Space

8.1   Sphere

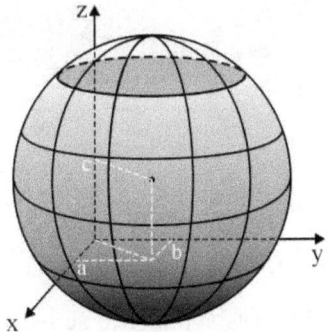

A sphere is the locus of all points in space for which the distance to point $M(a,b,c)$, or center is constant, and called radius.

From the formula of distance of any point (x,y,z) on the surface of the sphere, to point $M(a,b,c)$, the sphere equation is obtained:

$$(x-a)^2 + (y-b)^2 + (z-c)^2 = R^2$$

The intersection lines of a sphere with any plane are circles.
Example: The line of intersection of plane $z = h$, which is parallel to plane xy is the following circle

$$(x-a)^2 + (y-b)^2 = R^2 - (h-c)^2$$

when $|h-c| < R$.

8.2   Ellipsoid

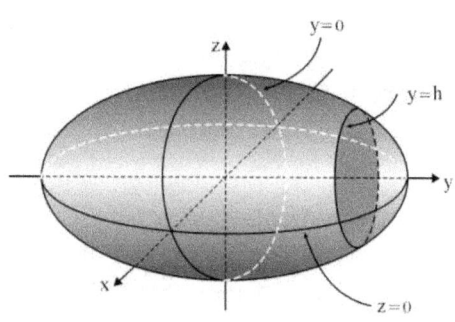

$$\frac{x^2}{a^2} + \frac{y^2}{b^2} + \frac{z^2}{c^2} = 1 \ , \ a > 0 \ , \ b > 0 \ , \ c > 0$$

is an ellipsoid that is symmetric across planes $x = 0$, $y = 0$, $z = 0$.

Intervals $[-a, a]$ on x-axis, $[-b, b]$ on y-axis and $[-c, c]$ on z axis are called ellipsoid axis. The lines of intersection of ellipsoid with planes parallel to coordinate planes are ellipses.

Example: $y = h$, when $|h| < b$, is

$$\frac{x^2}{a^2} + \frac{z^2}{c^2} = 1 - \frac{h^2}{b^2} .$$

Similarly, the lines of intersection with planes $z = h$ and $x = h$ are ellipses.

8.3   One-Sheet Hyperboloid

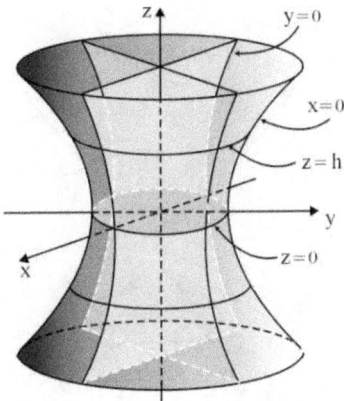

$$\frac{x^2}{a^2}+\frac{y^2}{b^2}-\frac{z^2}{c^2}=1$$

The hyperboloid is symmetric across coordinate planes $x=0$, $y=0$, $z=0$.

The intersection lines of one-sheet hyperboloid with the plane $z=h$ are ellipse $\frac{x^2}{a^2}+\frac{y^2}{b^2}=1+\frac{h^2}{c^2}$.

If $|h|$ is increasing, then the ellipses axes are increasing.

The smallest ellipse, when $h=0$, is called the **neck** of one-sheet hyperboloid.

The intersection lines of one-sheet hyperboloid with planes yz and xz are hyperbolas.

8.4   Two-Sheet Hyperboloid

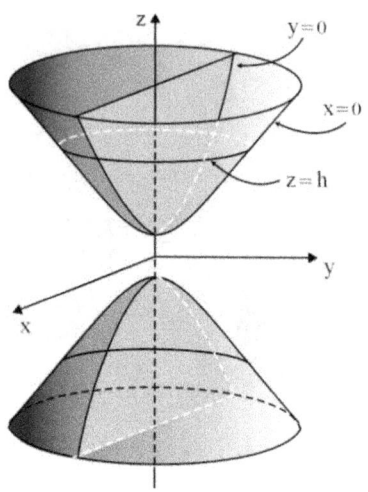

$$\frac{x^2}{a^2} + \frac{y^2}{b^2} - \frac{z^2}{c^2} = -1$$

The canonical form of two-sheet hyperboloid equation indicates it is symmetric across the planes of axes yz, xz, xy.

The intersection lines of two-sheet hyperboloid with planes $z = h$ are ellipses.

Plane $z = h$ starts intersecting two-sheet hyperboloid for $|h| \geq c$. In other words, between planes $z = c$ and $z = -c$, there exist no points belonging to this surface. The intersection lines with the planes at $x = h$ and $y = h$ are hyperbolas.

8.5   Cone

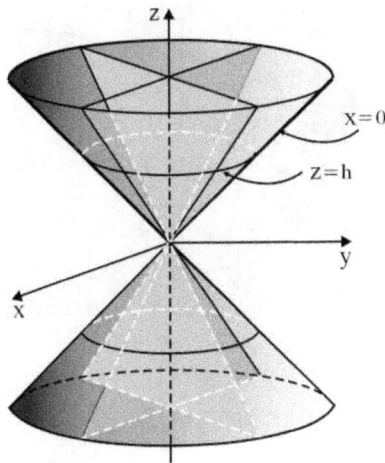

$$\frac{x^2}{a^2} + \frac{y^2}{b^2} - \frac{z^2}{c^2} = 0$$

The intersection lines of cone with planes $z = h$ are ellipses. The intersection line of the cone with plane $y = 0$ are a pair of straight lines $z = \pm \dfrac{c}{a} x$, and with plane $x = 0$, are a pair of straight lines $z = \pm \dfrac{c}{b} y$.

A cone is formed by straight lines passing through the origin.

8.6   Elliptic Paraboloid

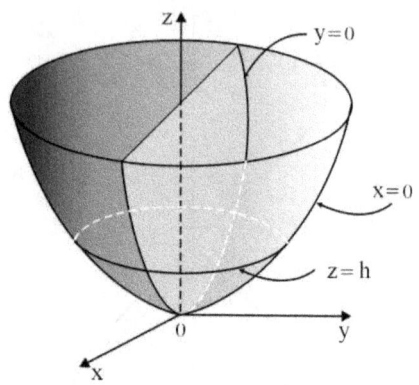

$$z = \frac{x^2}{a^2} + \frac{y^2}{b^2}$$

The intersection lines of elliptic paraboloid with planes $h > 0$, $z = h$ are ellipses $\dfrac{x^2}{a^2 h} + \dfrac{y^2}{b^2 h} = 1$ that increasingly widen with h increasing.

The intersection lines of it with planes $x = h$, $y = h$ are parabolas $z - \dfrac{h^2}{a^2} = \dfrac{y^2}{b}$ and $z - \dfrac{h^2}{b^2} = \dfrac{x^2}{a^2}$ respectively.

8.7   Hyperbolic Paraboloid

$$z = \frac{x^2}{a^2} - \frac{y^2}{b^2} \ , \ z = -\frac{x^2}{a^2} + \frac{y^2}{b^2}$$

Its intersection lines with planes $z = h$ are hyperbolas $\dfrac{x^2}{a^2 h} - \dfrac{y^2}{b^2 h} = 1$.

Its intersection lines with planes $x = 0$ and $y = 0$ are parabolas $z = \dfrac{y^2}{b^2}$ and $z = -\dfrac{x^2}{a^2}$ respectively.

Two straight lines pass through any point in a hyperbolic paraboloid.

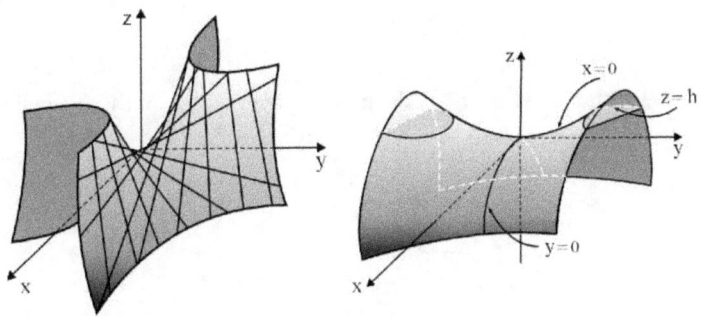

8.8   Cylindrical Surface

A **cylindrical surface** is formed by a straight line moving parallel to given straight line, along a certain curve. Any surface described by one of the equations $F(x, z) = 0$ (parallel to y-axis), $F(y, z) = 0$, (parallel to x-axis), $F(x, y) = 0$, (parallel to z-axis) is cylindrical. The following are examples of cylindrical surfaces:

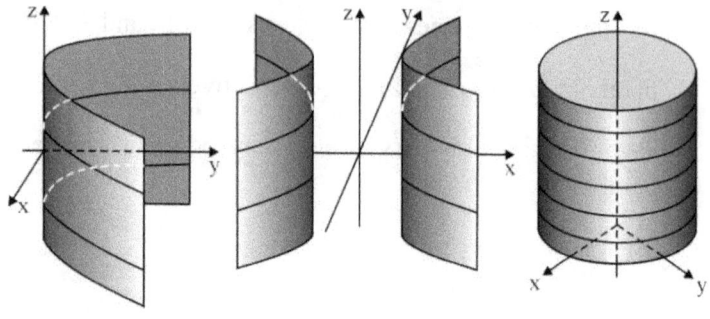

Parabolic cylinder	Hyperbolic cylinder	Elliptic cylinder
$x^2 = 2py, (p > 0)$	$\dfrac{x^2}{a^2} - \dfrac{y^2}{b^2} = 1$	$\dfrac{x^2}{a^2} + \dfrac{y^2}{b^2} = 1$

Chapter 9
Cylindrical and Spherical Coordinates

9.1   Cylindrical Coordinates

a. A **cylindrical coordinate** system combines the polar coordinates in a plane (see II 1.2), with vertical coordinate z in a space. The cylindrical coordinates of point M in space are ordered set of three numbers (ρ, φ, z), when (ρ, φ) are the polar coordinates of projection M_1 of point M on plane xy (see illustration). Remember that ρ is distance, and therefore can be either positive or zero, and angle φ varies from zero to 2π.

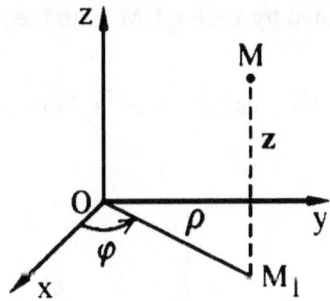

Formulas of conversion from cylindrical coordinates (ρ, φ, z) to Cartesian coordinates (x, y, z) are

$$x = \rho \cos \varphi \,,\ \ y = \rho \sin \varphi \,,\ \ z = z$$

To reverse the conversion, from (x, y, z) to (ρ, φ, z)

$$\rho = \sqrt{x^2 + y^2} \quad,\quad \tan \varphi = \frac{y}{x} \quad,\quad z = z$$

b. In cylindrical coordinates, the cylindrical surface equation has a simpler form.

Example: The equation of cylinder $x^2 + y^2 = a^2$ is

$$\rho = a,\ 0 \le \varphi < 2\pi,\ (a > 0)$$

That is, it is the equation of the locus for all sets of points at distance a from z-axis.

9.2   Spherical Coordinates

a. Point M in space is defined by the set of three numbers (ρ, φ, θ), when ρ is the length of vector $\overrightarrow{OM}$ (when O is the origin), φ is the angle of projection OM_1 with the positive direction of x-axis on plane xy, and θ is the angle formed by vector $\overrightarrow{OM}$ with the positive direction of z -axis.

(ρ,ϕ,θ) are called the **spherical coordinates** of point M .

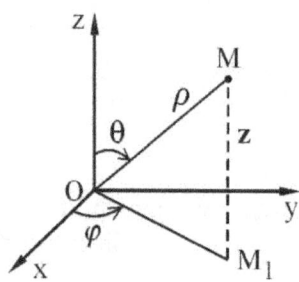

b. Relation between spherical and Cartesian coordinates:

1. If M is a point of spherical coordinates (ρ,ϕ,θ), then

$$x = \rho\cos\phi\cdot\sin\theta \ , \ y = \rho\sin\phi\cdot\sin\theta \ , \ z = \rho\cos\theta.$$

2. If M(x,y,z) are Cartesian coordinates of point M, then

$$\rho = \sqrt{x^2 + y^2 + z^2} \ , \ \cos\phi = \frac{x}{\sqrt{x^2 + y^2}} \ , \ \cos\theta = \frac{z}{\sqrt{x^2 + y^2 + z^2}}$$

when ρ only has positive values, ϕ values vary from zero to 2π and angle θ varies from zero to π .

c. **Examples**:

1. The spherical coordinates of point (2,1,2) are:

$$\rho = \sqrt{2^2 + 1 + 2^2} = 3 \ , \ \cos\phi = \frac{2}{\sqrt{5}} \Rightarrow \phi \approx 26.6° \ ,$$

$$\cos\theta = \frac{2}{3} \implies \theta \approx 48°$$

2. The equation of sphere $x^2 + y^2 + z^2 = R^2$ in spherical coordinates is

$$\rho = R \ , \ 0 \le \varphi \le 2\pi \ , \ -\frac{\pi}{2} \le \theta \le \frac{\pi}{2}$$

d. **Note**: Sometimes angle θ is referred to as the angle between vector $\overrightarrow{OM}$ and plane xy. In the illustration above, it is $\angle MOM_1$.

In such a case, θ is substituted with $\frac{\pi}{2} - \theta$, so the equation of sphere $x^2 + y^2 + z^2 = R^2$ in spherical coordinates is

$$\rho = R \ , \ 0 \le \varphi \le 2\pi \ , \ 0 \le \theta \le \pi.$$

IV. Single-Variable Differential Calculus

Chapter 1
Sequence, Limit of a Sequence

1.1   Definitions

A real numbers sequence is function $f : \mathbb{N} \to \mathbb{R}$.

$a_n = f(n), n \in \mathbb{N}$ is called the n-th member of sequence, or general term formula.

In other words, a sequence is an infinite ordered set of real numbers $a_1, a_2, \ldots, a_n, \ldots$, written in short as $\{a_n\}_{n=1}^{\infty}$.

Sequences $\{a_n\}_{n=1}^{\infty}$ and $\{b_n\}_{n=1}^{\infty}$ are equal if $a_n = b_n$ for all n.

Examples:

a. Arithmetic sequence is a sequence where each term, starting from the second one, equals the previous term plus a constant number d called the common difference.

$a_n = a_1 + (n-1)d$ is its general term formula, where a_1 is the first member.

$S_n = \dfrac{a_1 + a_n}{2} \cdot n$ is the n-th partial sum of the arithmetic sequence.

b. A geometric sequence is a sequence where each term, starting from the second one, equals the preceding term multiplied by a constant number q called the common ratio of the sequence.

$a_n = a_1 q^{n-1}$ is its general term formula.

$$S_n = \begin{cases} na_1 & ,q=1 \\ \dfrac{a_1(1-q^n)}{1-q} & ,q \neq 1 \end{cases}$$ is the n-th partial sum of the

sequence.

c. If, in an infinite geometric sequence, $|q|<1$, it has a finite sum:

$$a_1 + a_1 q + a_1 q^2 + \ldots + a_n q^n + \ldots = \frac{a}{1-q}$$

then the sum on the left side is called **converging geometric series** (see VI, 1.1).

d. Sequence $1, \dfrac{1}{2}, \dfrac{1}{3}, \dfrac{1}{4}, \ldots, \dfrac{1}{n}, \ldots$ is called **harmonic sequence**, and $a_n = \dfrac{1}{n}$ is the general term formula.

e. A **Leibniz sequence** is $1, -\dfrac{1}{2}, \dfrac{1}{3}, -\dfrac{1}{4}, \ldots, \dfrac{(-1)^{n+1}}{n}, \ldots$,

where $a_n = \dfrac{(-1)^{n+1}}{n}$ is the general term formula.

f. In a sequence such as $3,3,\ldots,3,\ldots$, consisting of one real number recurring infinitely, $a_n = 3$ is the general term.

1.2 Recursive Definition of a Sequence

a. The recursive definition of sequence has two parts:

1. The first term or first terms n value.

2. Recursion formula $a_{n+1} = f(a_1, a_2 ..., a_n)$ when function $f(x)$ is defined for all a_n.

b. **Example**

A given sequence is defined as:

$a_{n+1} = a_n + 3n^2 - 3n - 2$, $a_1 = -2$.

To find a_2, let's substitute $n = 1$, and the result will be $a_2 = a_1 + 3 \cdot 1^2 - 3 \cdot 1 - 2 = -4$.

Using a_2 to calculate a_3, we obtain

$a_3 = a_2 + 3 \cdot 2^2 - 3 \cdot 2 - 2 = 0$.

1.3   Bounded Sequences

a. Sequence $\{a_n\}_{n=1}^{\infty}$ is **bounded above** if there exists number M, called **upper bound**, such that $a_n \leq M$ for all natural n.

b. The smallest upper bound S of a sequence is called the sequence **supremum**, and is denoted as $S = \sup\{a_n\}$. If S is a term of the sequence, it is called **maximum**, and is denoted as $S = \max\{a_n\}$.

c. Sequence $\{a_n\}_{n=1}^{\infty}$ is called **bounded from below** if there exists number m, called lower bound, such that $a_n \geq m$ for all $n \in \mathbb{N}$.

d. Greatest lower bound I of a sequence is called **infimum**, and is denoted as $I = \inf\{a_n\}$. If it is a member of the sequence, it is called minimum and denoted as $I = \min\{a_n\}$.

e. Sequence $\{a_n\}_{n=1}^{\infty}$ is bounded if it is bounded from above and below. In the same manner, sequence $\{a_n\}$ is bounded if there exists number M such that for all natural n there holds $|a_n| \leq M$.

1.4   Increasing and Decreasing Sequences

a. Sequence $\{a_n\}_{n=1}^{\infty}$ is increasing if there exists n_0 such that for all $n \geq n_0$, $a_n \leq a_{n+1}$.

b. A sequence is strictly increasing if, for all $n \geq n_0$, $a_n < a_{n+1}$.

c. A sequence is decreasing if, for all $n \geq n_0$, $a_n \geq a_{n+1}$.

d. A sequence is strictly decreasing if, for all $n \geq n_0$, $a_n > a_{n+1}$.

e. An increasing/decreasing, or strictly increasing/decreasing, sequence is called a **monotonic** sequence.

Example: $\{n^2 + n - 5\}_{n=1}^{\infty}$ is a strictly increasing sequence.

Proof:

From $a_{n+1} - a_n = [(n+1)^2 + (n+1) - 5] - (n^2 + n - 5) = 2n + 2 > 0$ it follows that $a_{n+1} > a_n$ for all natural n.

1.5   Sub-Sequence

Sequence $\{b_k\}_{k=1}^{\infty}$ is a sub-sequence of sequence $\{a_n\}_{n=1}^{\infty}$ if all its terms are in sequence $\{a_n\}_{n=1}^{\infty}$ by the same order in which they appear $\{a_n\}$.

Examples:

a. Sequence $\left\{\dfrac{1}{n^2}\right\}_{n=1}^{\infty}$ is a sub-sequence of sequence $\left\{\dfrac{1}{n}\right\}_{n=1}^{\infty}$.

b. Sequence $\{4^n\}_{n=1}^{\infty}$ is a sub-sequence of sequence $\{2^n\}_{n=1}^{\infty}$.

1.6   Limit of a Sequence

a is the **limit of sequence** $\{a_n\}_{n=1}^{\infty}$ if, for every $\varepsilon > 0$ there exists natural number n_0 such that, for all $n \geq n_0$, there holds $|a_n - a| < \varepsilon$ or $a_n \in (a - \varepsilon, a + \varepsilon)$. In other words: for every given $\varepsilon > 0$, there is only a finite number of terms which are not in interval $(a - \varepsilon, a + \varepsilon)$. It is written $a = \lim\limits_{n \to \infty} a_n$ or

$a_n \xrightarrow[n \to \infty]{} a$.

If a sequence has a limit, it is called **convergent**.
If a series has no limit, it is called **divergent**.

1.7   Properties of Convergent Sequences

a. If a sequence has a limit, it is unique.

b. If a sequence converges to some limit, then every sub-sequence of it converges to the same limit.

c. If a sequence has two sub-sequences converging to two different limits, then the sequence diverges.

Example: The sequence $\left\{ \sin \dfrac{n\pi}{6} \right\}_{n=1}^{\infty}$ diverges because it has two sub-sequences converging to different limits:

If $n = 6k$, then sequence $\left\{ \sin \dfrac{6k\pi}{6} \right\}_{k=1}^{\infty} = \{0\}_{k=1}^{\infty}$ converges to zero.

If $n = 12k - 9$, then sequence $\left\{ \sin \dfrac{12k - 9}{6} \pi \right\}_{k=1}^{\infty} = \{1\}_{k=1}^{\infty}$ converges to 1.

d. Any converging sequence is bounded. The reverse proposition is incorrect.

Example: $1, -1, 1, -1, \ldots$ is bounded but not convergent sequence.

e. If sequences $\{a_n\}_{n=1}^{\infty}$ and $\{b_n\}_{n=1}^{\infty}$ converge, then:

1. For every constant c, $\lim\limits_{n \to \infty} c \cdot a_n = c \lim\limits_{n \to \infty} a_n$

2. $\lim\limits_{n \to \infty}(a_n \pm b_n) = \lim\limits_{n \to \infty} a_n \pm \lim\limits_{n \to \infty} b_n$

3. $\lim\limits_{n \to \infty}(a_n \cdot b_n) = (\lim\limits_{n \to \infty} a_n) \cdot (\lim\limits_{n \to \infty} b_n)$

4. If $\lim\limits_{n \to \infty} b_n \neq 0$ and $b_n \neq 0$ also, then $\lim\limits_{n \to \infty} \dfrac{a_n}{b_n} = \dfrac{\lim\limits_{n \to \infty} a_n}{\lim\limits_{n \to \infty} b_n}$.

f. If sequence $\{b_n\}_{n=1}^{\infty}$ is bounded and $\lim\limits_{n \to \infty} a_n = 0$, then $\lim\limits_{n \to \infty}(a_n b_n) = 0$.

Example: The sequence $\left\{\dfrac{1}{n}\right\}_{n=1}^{\infty}$ converges to zero, and sequence $\{\sin n\}_{n=1}^{\infty}$ is bounded by 1, therefore, $\lim\limits_{n \to \infty} \dfrac{\sin n}{n} = \lim\limits_{n \to \infty} \dfrac{1}{n} \cdot \sin n = 0$.

g. **Squeeze (sandwich) theorem**: if $\{a_n\}_{n=1}^{\infty}$, $\{b_n\}_{n=1}^{\infty}$, $\{c_n\}_{n=1}^{\infty}$, are three sequences where holds $\lim\limits_{n \to \infty} a_n = \lim\limits_{n \to \infty} c_n = a$ and $a_n \leq b_n \leq c_n$ for all n starting from n_0, then, $\lim\limits_{n \to \infty} b_n = a$.

h. If the sequence $\{a_n\}_{n=1}^{\infty}$ converges to a, and starting from n_0, $a_n \geq b$, then $a \geq b$.

i. **Cauchy's test**: real numbers sequence $\{a_n\}_{n=1}^{\infty}$ converges if, and only if,, for every $\varepsilon > 0$ there exists a natural number n_0 such that for all $n, m \geq n_0$ there holds $|a_m - a_n| < \varepsilon$.

1.8   Examples of Limits

a. $\lim\limits_{n\to\infty} \sqrt[n]{a} = 1$, $\lim\limits_{n\to\infty} \sqrt[n]{n} = 1$

b. $\lim\limits_{n\to\infty} \dfrac{1}{\sqrt[n]{n!}} = 0$, $\lim\limits_{n\to\infty} \dfrac{n}{\sqrt[n]{n!}} = e$

c. Let $\{a_n\}_{n=1}^{\infty}$ be a sequence of positive numbers.

If there exists a limit $\lim\limits_{n\to\infty} \dfrac{a_{n+1}}{a_n}$, then the sequence

$\{\sqrt[n]{a_n}\}_{n=1}^{\infty}$ converges and $\lim\limits_{n\to\infty} \sqrt[n]{a_n} = \lim\limits_{n\to\infty} \dfrac{a_{n+1}}{a_n}$.

d. If the sequence $\{a_n\}_{n=1}^{\infty}$ converges and $\lim\limits_{n\to\infty} a_n = a$, then:

1. $\lim\limits_{n\to\infty} |a_n| = |a|$

2. $\lim\limits_{n\to\infty} \sqrt{|a_n|} = \sqrt{|a|}$

3. $\lim\limits_{n\to\infty} \dfrac{a_1 + a_2 + \dots + a_n}{n} = a$, $\lim\limits_{n\to\infty} \sqrt[n]{a_1 \cdot a_2 \cdots a_n} = a$

4. $\lim\limits_{n\to\infty} \dfrac{n}{\dfrac{1}{a_1} + \dfrac{1}{a_2} + \dots + \dfrac{1}{a_n}} = a$

e. If $P(x) = a_0 + ax + a_2 x^2 + \dots + a_k x^k$

$Q(x) = b_0 + b_1 x + b_2 x^2 + \dots + b_m x^m$

are two polynomials, and suppose that $Q(n) \neq 0$ for all $n \in \mathbb{N}$, then

$$\lim_{n \to \infty} \frac{P(n)}{Q(n)} = \begin{cases} 0 & , k < m \\ \dfrac{a_k}{b_m} & , k = m \\ +\infty & , k > m , a_k \cdot b_m > 0 \\ -\infty & , k > m , a_k \cdot b_m < 0 \end{cases}.$$

1.9   Convergence to Infinity

a. Sequence $\{a_n\}_{n=1}^{\infty}$ tends to infinity or minus infinity, if for every real number M there only exists a finite number of terms smaller or greater than M. In other words, for every M there exists $n_0 \in \mathbb{N}$ such that for all $n \geq n_0$ holds $a_n > M$ or $a_n < M$.

It is written $\lim_{n \to \infty} a_n = \infty$ or $\lim_{n \to \infty} a_n = -\infty$.

b. A sequence **converges in the broad sense** if it converges in the ordinary sense or it tends to infinity or minus infinity.

1.10   Limits of Monotonic Sequences. The e Number

a. Every monotonic and bounded sequence, either increasing or decreasing, converges. Every monotonic sequence converges in the broad sense.

b. Every increasing, unbounded sequence tends to infinity.

c. Every decreasing, unbounded sequence tends to negative infinity.

d. The sequence $\{a_n\}_{n=1}^{\infty} = \left\{ \left(1 + \dfrac{1}{n}\right)^n \right\}_{n=1}^{\infty}$ is increasing and bounded, and $2 < a_n < 3$ and therefore, it had a finite limit denoted as e

$$\lim_{n \to \infty}\left(1+\frac{1}{n}\right)^n = e \approx 2.71828182845\ldots$$

1.11 Cantor Theorem

Let $[a_n, b_n]$ be a sequence of closed intervals such that, for all $n \in \mathbb{N}$, holds

$$[a_{n+1}, b_{n+1}] \subseteq [a_n, b_n],\ \lim_{n \to \infty}(b_n - a_n) = 0 .$$

Then, there is only one point c belonging to all intervals $[a_n, b_n]$.

1.12 Bolzano-Weierstrass Theorem

Every bounded sequence has a converging sub-sequence.

Chapter 2
Limits of Functions

2.1   Definitions of Limit

a. Heine's Definition of Limit

1. Let function $f(x)$ be defined in the neighborhood of point $x = a$, except, possibly, point a itself. A real number L is the limit of $f(x)$ when $x \to a$ if, for all sequence $\{x_n\}_{n=1}^{\infty}$ converging to limit a, such that $x_n \neq a$, the sequence $\{f(x_n)\}_{n=1}^{\infty}$ converges to L.

2. Let function $f(x)$, be defined in ray (α, ∞). A real number L is the limit of $f(x)$ when x tends to

infinity, if for all sequence $\{x_n\}_{n=1}^{\infty}$ converging to infinity, the sequence $\{f(x_n)\}_{n=1}^{\infty}$ converges to L.

b. Cauchy's Definition of Limit

1. Let function $f(x)$ be defined in the neighborhood of point $x = a$, except, possibly, point a itself. L is the limit of $f(x)$ when $x \rightarrow a$, $(x \neq a)$, if for every arbitrary small number $\varepsilon > 0$ there exists a $\delta > 0$ such that for all x which holds $0 < |x - a| < \delta$, there holds $|f(x) - L| < \varepsilon$, or, in logic symbols

$$\lim_{x \to a} f(x) = L \Leftrightarrow$$

$$\forall \varepsilon > 0, \exists \delta > 0, \forall x : 0 < |x - a| < \delta \Rightarrow |f(x) - L| < \varepsilon$$

2. L is the limit of $f(x)$ when x tends to infinity, if, for an arbitrary small $\varepsilon > 0$ there exists E such that for all x holding $x > E$, there holds $|f(x) - L| < \varepsilon$ or, in logic symbols

$$\lim_{x \to \infty} f(x) = L \Leftrightarrow \forall \varepsilon > 0 , \exists E , x > E \Rightarrow |f(x) - L| < \varepsilon.$$

c. The limit of $f(x)$ at point $x = a$, is infinity $(\lim_{x \to a} f(x) = \infty)$, if for all $M > 0$ there exists $\delta > 0$ such that for all x that holds $0 < |x - a| < \delta$, there holds $f(x) > M$.

d. Heine's and Cauchy's definitions are equivalent.

2.2   Properties of Limits

a. Let $f(x)$ and $g(x)$ be two functions, and $\lim_{x \to a} f(x) = A$, $\lim_{x \to a} g(x) = B$ are their limits. Then:

1. $\lim_{x \to a} c = c$, for every constant c

2. $\lim\limits_{x \to a} c f(x) = c \lim\limits_{x \to a} f(x) = cA$

3. $\lim\limits_{x \to a}[f(x) \pm g(x)] = \lim\limits_{x \to a} f(x) \pm \lim\limits_{x \to a} g(x) = A \pm B$

4. $\lim\limits_{x \to a}[f(x) \cdot g(x)] = [\lim\limits_{x \to a} f(x)][(\lim\limits_{x \to a} g(x)] = A \cdot B$

5. If $B \neq 0$, then $\lim\limits_{x \to a} \dfrac{f(x)}{g(x)} = \dfrac{\lim\limits_{x \to a} f(x)}{\lim\limits_{x \to a} g(x)} = \dfrac{A}{B}$

b. If function $f(x)$ is bounded and $\lim\limits_{x \to a} g(x) = 0$, then $\lim\limits_{x \to a} f(x) \cdot g(x) = 0$.

Example: $|\cos\dfrac{1}{x}| \leq 1$ and $\lim\limits_{x \to 0} x^2 = 0$.

Therefore, $\lim\limits_{x \to 0} x^2 \cos\dfrac{1}{x} = 0$.

c. **Squeeze (sandwich) theorem**: Let $f(x)$, $g(x)$, $h(x)$ be three functions defined in a neighborhood of point $x = a$, except, possibly, point a itself. If, for all x in this neighborhood, there holds $f(x) \leq g(x) \leq h(x)$ and the limits $A = \lim\limits_{x \to a} f(x) = \lim\limits_{x \to a} h(x)$ exists, then, $\lim\limits_{x \to a} g(x) = A$.

d. If function $f(x)$ has a limit, then it is unique.

e. If there exist two different sequences $\{x_n\}_{n=1}^{\infty}$, $\{y_n\}_{n=1}^{\infty}$ converging to a, but sequences $\{f(x_n)\}_{n=1}^{\infty}$ and $\{f(y_n)\}_{n=1}^{\infty}$ converges to different limit, then, there is no limit for $f(x)$ when $x \to a$.

Example: For function $f(x) = e^{x-[x]}$, when [x] is the integer value of x, there exists no limit when $x \to \infty$.

Proof: Let's take the sequence $x_n = n$, holding $\lim\limits_{x\to\infty} x_n = \infty$, and, for all n, $x_n - [x_n] = n - n = 0$. Therefore, $f(x_n) = e^0 = 1$, and therefore $\lim\limits_{x\to\infty} f(x_n) = 1$.

Now, let's take the sequence $y_n = n + \dfrac{1}{2}$. Here, $\lim\limits_{n\to\infty} y_n = \infty$. For all n, $y_n - [y_n] = \dfrac{1}{2}$, and $\lim\limits_{n\to\infty} f(y_n) = \lim\limits_{n\to\infty} e^{\frac{1}{2}} = \sqrt{e}$. That is, there doesn't exist a limit $\lim\limits_{x\to\infty} f(x)$.

f. If $f(x)$ is an elementary function, (see, I, 1.9) defined at point x_0, then $\lim\limits_{x\to x_0} f(x) = f(x_0)$.

g. $\lim\limits_{x\to 0} \dfrac{\sin x}{x} = 1$.

h. $\lim\limits_{x\to\infty}\left(1 + \dfrac{1}{x}\right)^x = e$, $\lim\limits_{\alpha\to 0}(1+\alpha)^{\frac{1}{\alpha}} = e$.

i. If function $f(x)$ is defined in the neighborhood of $x = a$ and there exists a limit $\lim\limits_{x\to a} f(x) = L$, then:

   1. There exists a neighborhood of a where $f(x)$ is bounded.

   2. If $L > 0$, then there exists a neighborhood of a, such that for all x in this neighborhood, (except, possibly, a), where there holds $f(x) > 0$.

   3. If $L < 0$, then there exists a neighborhood of a, such that for all x in this neighborhood, (except, possibly, a), where there holds $f(x) < 0$.

2.3   One-Sided Limits

a. Let $f(x)$ be a function defined in the right-handed neighborhood of point $x = a$. That is, let there exist $r > a$ such that $f(x)$ is defined in interval (a, r). Real number L is the **right-handed limit** of $f(x)$ when x tends to point a to the right $x \to a^+$, (always $x > a$), if, for every $\varepsilon > 0$, there exists $\delta > 0$ such that

$$a < x < a + \delta \implies |f(x) - L| < \varepsilon.$$

Let's write $\lim\limits_{x \to a^+} f(x) = L$.

b. Let $f(x)$ be a function defined in the left-handed neighborhood of point $x = a$. That is, let there exist $r < a$ such that $f(x)$ is defined in interval (r, a). Real number L is the **left-handed limit** of $f(x)$ when $x \to a^-$, if for every $\varepsilon > 0$ there exists $\delta > 0$ such that holds $a - \delta < x < a \implies |f(x) - L| < \varepsilon$. Let's write $\lim\limits_{x \to a^-} f(x) = L$.

Example: $\lim\limits_{x \to 1^-} 2^{\frac{1}{1-x}} = \infty$, $\lim\limits_{x \to 1^+} 2^{\frac{1}{1-x}} = 0$.

c. For function $f(x)$, there exists a limit at point $x = a$ if, and only if, there exist one-sided limits at this point, and $\lim\limits_{x \to a^+} f(x) = \lim\limits_{x \to a^-} f(x) = \lim\limits_{x \to a} f(x)$ holds.

Chapter 3
Continuity of Functions

Function $f(x)$ is continuous at point x_0 if it is defined in the neighborhood of this point, there exists the limit $\lim\limits_{x \to x_0} f(x)$, and it equals to the value of the function at point x_0. That is, $\lim\limits_{x \to x_0} f(x) = f(x_0)$.

3.1   Properties of Continuous Functions

a. If an elementary function is continuous at point x_0, then it is continuous at x_0.

b. If functions $f(x)$ and $g(x)$ are defined at point x_0, then:

1. For every constant c, function $cf(x)$ is continuous at x_0.

2. $f(x) \pm g(x)$ is continuous at x_0.

3. $f(x)g(x)$ is continuous at x_0.

4. If, in addition, $g(x_0) \neq 0$, then $\dfrac{f(x)}{g(x)}$ is continuous at x_0.

c. The composition of two continuous functions is a continuous function.

d. Let $f(x)$ be a function continuous on a closed interval $[a,b]$. If $f(a) \cdot f(b) < 0$ (that is, $f(x)$) attains values of opposite signs at the ends points of the interval, then there exists a point $a < c < b$, such that $f(c) = 0$.

e. For all continuous function $f:[a,b]\rightarrow[a,b]$, there exists a point $a\leq c\leq b$, such that $f(c)=c$. It is called **stationary point**.

f. **Cauchy's intermediate value theorem**: If function $f(x)$ is continuous in interval $[a,b]$, and real number m is between $f(a)$ and $f(b)$, then there exists a point x_0, $a<x_0<b$, such that $m=f(x_0)$.

g. **Weierstrass extreme value theorem**: If function $f(x)$ is continuous in finite, closed interval $[a,b]$, then $f(x)$ is bounded in that interval, and attains its maximum and minimum in that interval. That is, there exists an $\underline{x}$ in the interval such that $f(\underline{x})\leq f(x)$, for all $a\leq x\leq b$, and there exists an $\overline{x}$ in the interval such that $f(x)\leq f(\overline{x})$ for all $a\leq x\leq b$.

3.2   Types of Discontinuities

a. Let $f(x)$ be a function defined in the neighborhood of $x=x_0$, except, possibly, x_0. We say that x_0 is **removable discontinuity** if:

1. There exists a limit $\lim_{x\to x_0}f(x)$.

2. $\lim_{x\to x_0}f(x)\neq f(x_0)$, or the function is undefined at x_0.

Example: For function $f(x)=\dfrac{x^2-4}{x-2}$, $x=2$ is a removable discontinuity point since there exists $\lim_{x\to 2}\dfrac{x^2-4}{x-2}=4$, and the function is undefined at $x=2$.

b. Point $x = x_0$ is called **first type** or **jump discontinuity** of $f(x)$, if:

1. $f(x)$ is defined in a specific neighborhood of $x = x_0$, except, possibly, x_0.

2. It has the one-sided, finite limits $\lim\limits_{x \to x_0^-} f(x)$, $\lim\limits_{x \to x_0^+} f(x)$.

3. $\lim\limits_{x \to x_0^-} f(x) \neq \lim\limits_{x \to x_0^+} f(x)$.

Example: For function $f(x) = \dfrac{|x|}{x}$, $x = 0$ is a jump discontinuity since $f(x)$ is defined in the neighborhood of $x = 0$ (except in $x = 0$), and the two one-sided limits $\lim\limits_{x \to 0^+} \dfrac{|x|}{x} = 1$, $\lim\limits_{x \to 0^-} \dfrac{|x|}{x} = -1$ exist and they are different.

c. x_0 is called second type discontinuity of $f(x)$ if:

1. $f(x)$ is defined in the neighborhood of x_0, except, possibly, x_0.

2. At list one of the one-sided limits at x_0 does not exists or in not finite.

Examples:

1. For function $f(x) = e^{\frac{1}{x}}$ has a second-type discontinuity at $x = 0$, since $\lim\limits_{x \to 0^+} e^{\frac{1}{x}} = \infty$.

2. For function $f(x) = \cos\dfrac{1}{x}$, $x = 0$ is a second-type discontinuity. $\lim\limits_{x \to 0^+} \cos\dfrac{1}{x}$ does not exist.

3.3   Uniform Continuity

Let function $f(x)$ be defined above interval I. We say that $f(x)$ is uniformly continuous in I, if, for every $\varepsilon > 0$ there exists $\delta > 0$ such that for all x_1, x_2 for which holds $|x_1 - x_2| < \delta$, there holds

$$|f(x_1) - f(x_2)| < \varepsilon.$$

Cantor theorem: A function continuous at closed interval $[a, b]$, is uniformly continuous in $[a, b]$.

Chapter 4
Derivative

4.1   Definition

Let:

$f(x)$ be a function defined in the neighborhood of $x = x_0$.

Δx be a numeric variable added to x_0 (it can be either positive or negative).

$x = x_0 + \Delta x$ **points of the neighborhood of** x_0

Function $f(x)$ is called differentiable at point $x = x_0$, if there exists a finite limit

$$L = \lim_{\Delta x \to 0} \frac{\Delta y}{\Delta x} = \lim_{\Delta x \to 0} \frac{f(x_0 + \Delta x) - f(x_0)}{\Delta x}.$$

Number L is the functions' derivative at point $x = x_0$.

Other denotations of derivative: $f'(x_0)$, $\dfrac{dy}{dx}$, $\dfrac{df}{dx}$, y', y'_x.

If we substitute Δx with the expression $x - x_0$, which equals it, we could also write the definition of derivative at $x = x_0$ the following way:

$$f'(x_0) = \lim_{\Delta x \to 0} \frac{f(x_0 + \Delta x) - f(x_0)}{\Delta x} = \lim_{x \to x_0} \frac{f(x) - f(x_0)}{x - x_0}.$$

4.2   Tangent Line to a Curve: Geometric Description of the n Derivative

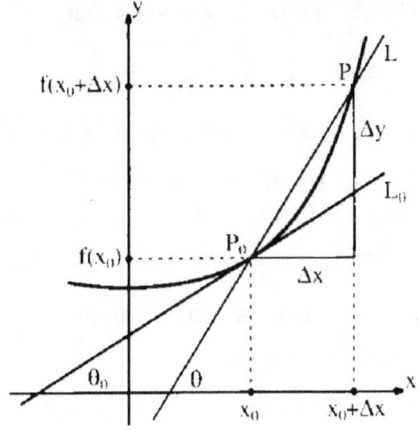

a. A **straight line tangent** to the graph of function $y = f(x)$, if it exists, is the line obtained as the limit of straight lines $L = PP_0$ when point P tends to P_0 (see illustration).

Equivalently, the tangent straight line L_0 to the graph at x_0, is the limit of straight line L when $\Delta x = x - x_0$ tends to zero.

b. $f'(x_0)$ is the **slope** of the straight line tangent to the graph of the function at x_0, and there holds $f'(x_0) = tg\theta_0$ when θ_0 is the angle between the tangent line and the positive direction of the x-axis.

c. The tangent equation is $y = f(x_0) + f'(x_0) \cdot (x - x_0)$.

4.3   Linear Approximation

a. Function $f(x)$, defined at a specific neighborhood of point $x = x_0$ is differentiable at x_0 if, and only if, there exists a constant A and there exists function $\alpha(\Delta x)$, for which holds $\lim\limits_{\Delta x \to 0} \alpha(\Delta x) = 0$ such that

$$\Delta y = f(x_0 + \Delta x) - f(x_0) = A \cdot \Delta x + \alpha(\Delta x) \cdot \Delta x.$$

b. If $f(x)$ is differentiable, then $A = f'(x_0)$ for a small enough Δx, we write $f(x) \approx f(x_0) + f'(x_0) \cdot \Delta x$. That is, the graph of function $f(x)$ can be approximated to a straight line in a small enough neighborhood of x_0.

c. If function $f(x)$ is differentiable at x_0, then it is continuous at this point.

d. Not any function continuous at x_0 is differentiable at x_0.

   Example: $f(x) = |x|$ is continuous at $x = 0$ but not differentiable at $x = 0$.

4.4   Derivative Rules

Let $f(x), g(x), h(x)$ be derivable functions, and c a constant. Therefore:

a. $(cf(x))' = cf'(x)$

b. $[f(x) \pm g(x)]' = f'(x) \pm g'(x)$

c. $[f(x) \cdot g(x)]' = f'(x) \cdot g(x) + f(x) \cdot g'(x)$

d. $[f(x) \cdot g(x) \cdot h(x)]' = f'(x) \cdot g(x) \cdot h(x) +$

$+f(x) \cdot g'(x) \cdot h(x) + f(x) \cdot g(x) \cdot h'(x)$

e. $\left[\dfrac{f(x)}{g(x)}\right] = \dfrac{f'(x) \cdot g(x) - f(x) \cdot g'(x)}{[g(x)]^2}$, $g(x) \neq 0$

f. **Chain Rule: Derivative Composite Functions**

If $y = f(x)$ is function derivable at $x = x_0$, and $z = g(y)$ is function derivable at $y = y_0 = f(x_0)$, then the composite function $z = g(f(x))$ is derivable at x_0, and there holds the equality:

$$\frac{dz}{dx} = [g(f(x_0))]' = g'(y_0) \cdot f'(x_0) = \frac{dz}{dy} \cdot \frac{dy}{dx}$$

In other words, the derivative of composite function $g(f(x))$ equals to the derivative of $g(y)$ multiplied by the derivative of its inner function $f(x)$.

g. **Derivative Inverse Function**

If $y = f(x)$ is invertible function in the neighborhood of x_0, differentiable at x_0, and $f'(x_0) \neq 0$, then is inverse function, $x = g(y)$, is differentiable at $y_0 = f(x_0)$, and there holds

$$g'(y_0) = \frac{1}{f'(x_0)}$$

h. Derivative functions in the form of $y = [f(x)]^{g(x)}$ when $f(x) > 0$:

Using the logarithm function $\ln y = g(x) \cdot \ln f(x)$, we differentiate:

$$\frac{y'}{y} = g'(x) \cdot \ln f(x) + \frac{g(x) \cdot f'(x)}{f(x)}$$

$$y' = [f(x)]^{g(x)} \left[g'(x) \cdot \ln f(x) + \frac{g(x) \cdot f'(x)}{f(x)} \right]$$

Using another way: $y = [f(x)]^{g(x)} = e^{g(x)\ln f(x)}$ and deriving.

i. **Deriving a Function Presented in its Parametric Form:**

Let $y = f(x)$ be a function given in the parametric form of $x = \varphi(t)$, $y = \psi(t)$ (see II.3), when $\varphi(t)$ is invertible in the given domain, that is $t = \varphi^{-1}(x)$, and therefore, $y = \psi(\varphi^{-1}(x))$. Then, from chain rule, there follows:

$$\frac{dy}{dx} = \frac{dy}{dt} \cdot \frac{dt}{dx} = \frac{\psi'(t)}{\varphi'(t)}$$

4.5   Derivatives of Elementary Functions

	$f(x)$	$f'(x)$		$f(x)$	$f'(x)$
1.	c	0	9.	e^x	e^x
2.	x	1	10.	$\ln\lvert x\rvert$	$\dfrac{1}{x}$
3.	x^α	$\alpha x^{\alpha-1}$	11.	$\arcsin x$	$\dfrac{1}{\sqrt{1-x^2}}$
4.	$\sin x$	$\cos x$	12.	$\arccos x$	$-\dfrac{1}{\sqrt{1-x^2}}$

5.	cos x	$-\sin x$	13.	arctan x	$\dfrac{1}{1+x^2}$
6.	tan x	$\dfrac{1}{\cos^2 x}$	14.	arc cot x	$-\dfrac{1}{1+x^2}$
7.	cot x	$-\dfrac{1}{\sin^2 x}$	15.	sinh x	cosh x
8.	a^x	$a^x \ln a$	16.	cosh x	sinh x

4.6   One-Sided Derivatives

a. If function $f(x)$ is defined in a right-handed neighborhood of $x = x_0$, and there exists limit
$$\lim_{\Delta x \to 0^+} \frac{f(x_0 + \Delta x) - f(x_0)}{\Delta x},$$ it is called the **right-hand derivative** of $f(x)$ at x_0, and is denoted by $f'_+(x_0)$.

b. The **left-handed derivative** of $f(x)$ at x_0 is
$$f'_-(x_0) = \lim_{\Delta x \to 0^-} \frac{f(x_0 + \Delta x) - f(x_0)}{\Delta x}.$$

c. **Example**: The one-sided derivatives of $f(x) = |x|$ are
$$f'_+(0) = 1 \, , \ f'_-(0) = -1.$$

d. The function $f(x)$ is differentiable at x_0 if, and only if, its one-sided derivatives at x_0 exist are equal.

e. If the derivative at x_0 exists, then
$$f'(x_0) = f'_+(x_0) = f'_-(x_0).$$

f. **Function** $f(x)$ is **differentiable in close interval** $[a,b]$, if it is differentiable at (a,b) and its one-sided derivatives $f'_+(a)$, $f'_-(b)$ exist.

4.7   High-Order Derivatives

a. Let $y = f(x)$ be a function derivable at interval (a,b). If we derivative it at all points of the interval, we'll have a new function $f'(x)$, which, too, is defined in all interval (a,b). The function $y' = f'(x)$ is the **first derivative** of $f(x)$ in interval (a,b).

If function $f'(x)$, also, is derivable in interval (a,b), we will denote its derivative as $f''(x)$. $y'' = f''(x)$ is defined on all interval (a,b), and is called the **second derivative** of $f(x)$. Similarly, we define the third derivative, forth derivative, and so on.

In general, if function $f(x)$ can be derivated n times, in interval (a,b), then, the last function in the set of n derivatives is denoted by $f^{(n)}(x)$ or by $y^{(n)}$, and is called the n **-th derivative** of $f(x)$.

Another denotation of n -th derivative is that of Leibniz: $\dfrac{d^n f}{dx^n}$.

b. **Leibniz formula**

Let $u(x)$ and $v(x)$ be functions differentiable n times at x. Then:

$(uv)' = u'v + uv'$

$(uv)'' = u''v + 2u'v' + uv''$

$(uv)''' = u'''v + 3u''v' + 3u'v'' + uv'''$

$$[u(x)v(x)]^{(n)} = \sum_{k=0}^{n} \binom{n}{k} u^{(n-k)}(x) \cdot v^{(k)}(x)$$

$$\text{when } \binom{n}{k} = \frac{n!}{k!(n-k)!} \, .$$

c. **Examples**:

1. $(x^{\alpha})^{(n)} = \alpha(\alpha-1)\ldots(\alpha-n+1)x^{\alpha-n}$

2. $(\ln x)^{(n)} = (-1)^{n-1} \cdot (n-1)! x^{-n}$

3. $(e^x)^{(n)} = e^x$, $(a^x)^{(n)} = a^x \ln^n a$

4. $(\sin x)^{(n)} = \sin\left(x + \dfrac{\pi n}{2}\right)$, $(\cos x)^{(n)} = \cos\left(x + \dfrac{\pi n}{2}\right)$

4.8   Differentials

a.

1. The entity $dy = f'(x_0)dx = f'(x_0)\Delta x$ is called the **differential** of $f(x)$ at point x_0. As opposed to dx variable, which does not depend on any other entity, the dy variable depends on dx and x_0.

Another basic essential difference between dx and dy is that while $dx = \Delta x$, usually, $dy \neq \Delta y$. The illustration shows the geometric description of each of these entities, and the way the Δy entity is related to dy entity. These two entities tend to unite as dx decreases.

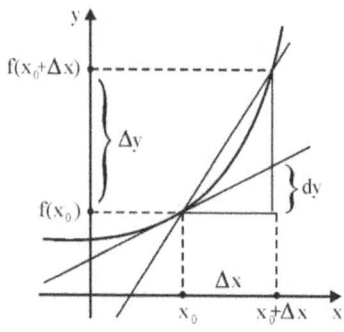

2. If $f'(x_0) \neq 0$, then $\lim\limits_{\Delta x \to 0} \dfrac{\Delta y}{dy} = 1$.

b. Higher-Order Differentials

$d^2 y = d(dy) = f''(x_0)(dx)^2$ is a second-order differential.

$d^n y = f^{(n)}(x_0)(dx)^n$ is an n-th order differential.

4.9 Differential Calculus Basic Theorems

a.

1. Fermat's Theorem

Let $f(x)$ be a function defined on open interval (a,b) and differentiable at inner point x_0. If $f(x)$ attains a maximum or a minimum value at x_0, then $f'(x_0) = 0$.

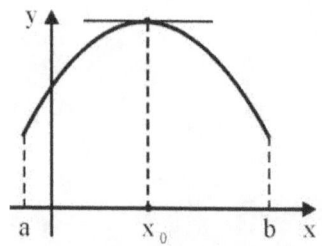

2. The Geometric Meaning of Fermat's Theorem

If $f(x)$ attains a maximum or a minimum value at x_0, and if, in the neighborhood of x_0, its graph has a "hill," then, the straight line tangent to $f(x)$ at this point have to parallel to the x-axis. That is, the derivative at x_0 should be zero.

b. Rolle's Theorem

Let $f(x)$ be a function defined at close interval $[a,b]$, and there holds the following:

1. $f(x)$ is continuous at close interval $[a,b]$.

2. $f(x)$ is differentiable at open interval (a,b).

3. $f(a) = f(b)$.

Then, there is a $c \in (a,b)$, such that $f'(c) = 0$.

Geometrically, Rolle's theorem means that if a continuous and differentiable function at a close interval has equal values at the extreme points of the interval, then there exists at least one point within the interval where the straight line tangent to the graph is parallel to the x-axis. But sometimes, there is more than one such point.

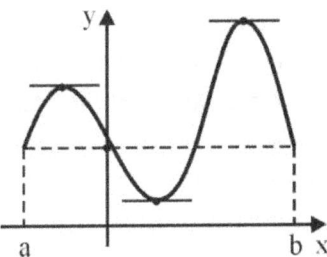

c. **Lagrange's Mean-Value Theorem**

1. If $f(x)$ is a function continuous at close interval $[a,b]$ and differentiable at open interval (a,b), then there is at least one point c, $a<c<b$ such that

$$f'(c) = \frac{f(b)-f(a)}{b-a}$$

2. Geometrically, the theorem means that if $f(x)$ is a function continuous at close interval $[a,b]$ and differentiable at open interval (a,b), then there exists a point on the graph where the straight line tangent to the graph in it is parallel to the straight line passing through the extreme points of the graph. The illustration shows the graph of a function with three such points.

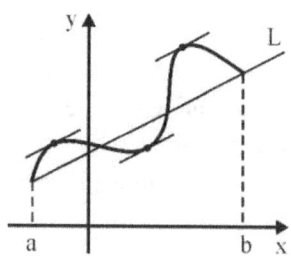

3. Another form of Lagrange's theorem: if $f(x)$ is a function differentiable at interval $(x_0-\delta,x_0+\delta)$, then, for all Δx, such that $x_0+\Delta x \in (x_0-\delta,x_0+\delta)$, there exists a real number θ, $0<\theta<1$, such that

$$f(x_0+\Delta x)-f(x_0) = f'(x_0+\theta\Delta x)\cdot \Delta x.$$

d. Cauchy's Mean-Value Theorem

1. Let $f(x)$ and $g(x)$ be two functions continuous at close interval $[a,b]$ and differentiable at open interval (a,b), and, in addition, $g'(x) \neq 0$ for all $a < x < b$. Then, there is at least one point c, $a < c < b$, such that

$$\frac{f(b)-f(a)}{g(b)-g(a)} = \frac{f'(c)}{g'(c)}$$

2. If $f(x)$ and $g(x)$ are functions continuous and differentiable in the neighborhood of x_0, and, $g'(x) \neq 0$ for all x in that neighborhood, then, for every Δx, (small enough) such that $x_0 + \Delta x$ is in the given neighborhood, there exists a real number θ, $0 < \theta < 1$, such that

$$\frac{f(x_0 + \Delta x) - f(x_0)}{g(x_0 + \Delta x) - g(x_0)} = \frac{f'(x_0 + \theta \Delta x)}{g'(x_0 + \theta \Delta x)}$$

e. Darboux's Mean-Value Theorem

1. If function $f(x)$ is differentiable at close interval $[a,b]$, then for every β between $f'_+(a)$ and $f'_-(b)$ there exists an $a \leq x_0 \leq b$ such that $f'(x_0) = \beta$. In other words, if $f(x)$ is differentiable at the close interval $[a,b]$, then, the image of $f'(x)$ is an interval.

2. If function $f(x)$ is differentiable at close interval $[a,b]$, then its derivative $f'(x)$ is not necessarily continuous and therefore is not differentiable.

If $f'(x)$ is not continuous at x_0, then it is a second-type discontinuity.

Example: The derivative of $f(x) = \begin{cases} x^2 \sin\dfrac{1}{x}, & (x \neq 0) \\ 0, & (x = 0) \end{cases}$ is

function $f'(x) = \begin{cases} 2x\sin\dfrac{1}{x} - \cos\dfrac{1}{x}, & (x \neq 0) \\ 0, & (x = 0) \end{cases}$, which is not

continuous at $x = 0$, since the limit $\lim\limits_{x \to 0} \cos\dfrac{1}{x}$ does not exist.

4.10 L'Hopital's Rules

a. Let $f(x)$ and $g(x)$ be functions differentiable in the neighborhood of $x = a$, except, possibly, at a. Suppose that:

1. There exists the limit $\lim\limits_{x \to a} f(x) = \lim\limits_{x \to a} g(x) = 0$

or $\lim\limits_{x \to a} f(x) = \lim\limits_{x \to a} g(x) = \infty$.

2. $g'(x) \neq 0$ for all $x \neq a$ in the neighborhood of a.

3. There exists the limit $\lim\limits_{x \to a} \dfrac{f'(x)}{g'(x)}$.

Then, there also exists the limit $\lim\limits_{x \to a} \dfrac{f(x)}{g(x)}$, and there holds that $\lim\limits_{x \to a} \dfrac{f(x)}{g(x)} = \lim\limits_{x \to a} \dfrac{f'(x)}{g'(x)}$.

b. If functions $f(x)$ and $g(x)$ are differentiable in infinite interval (a, ∞), and

1. The limits $\lim\limits_{x \to \infty} f(x) = \lim\limits_{x \to \infty} g(x) = 0$ exist

2. $g'(x) \neq 0$ for all $x > \alpha$

3. The limit $\lim\limits_{x \to \infty} \dfrac{f'(x)}{g'(x)}$ exists.

Then, the limit $\lim\limits_{x \to \infty} \dfrac{f(x)}{g(x)}$ also exists, and there holds

$$\lim_{x \to \infty} \frac{f(x)}{g(x)} = \lim_{x \to \infty} \frac{f'(x)}{g'(x)}.$$

4.11 Taylor's Formula

a. If function $f(x)$ is differentiable $n+1$ times in the neighborhood of $x = a$, and x is a point in this neighborhood, then there exists a point c, between a and x, such that

$$f(x) = f(a) + \frac{f'(a)}{1!}(x-a) + \frac{f''(a)}{2!}(x-a)^2 + \dots$$

$$+ \frac{f^{(n)}(a)}{n!}(x-a)^n + R_n(x)$$

when $R_n(x) = \dfrac{f^{(n+1)}(c)}{(n+1)!}(x-a)^{n+1}$ is **Lagrange remainder**.

If, we substitute $a = 0$ in Taylor's formula, we obtain **Maclaurin formula**:

$$f(x) = f(0) + \frac{f'(0)}{1!}x + \frac{f''(0)}{2!}x^2 + \dots + \frac{f^{(n)}(0)}{n!}x^n + \frac{f^{(n+1)}(c)}{(n+1)!}x^{n+1}$$

b. **Peano Remainder Formula**

$$R_n(x) = o\big((x-a)^n\big) , \quad x \to a$$

c. **Cauchy Remainder Formula**

$$R_n(x) = \frac{f^{(n+1)}\big(a + \theta(x-a)\big)}{n!}(1-\theta)^n (x-a)^{n+1}, \ 0 < \theta < 1$$

d. **Examples**:

1. $e^x = 1 + \dfrac{x}{1!} + \dfrac{x^2}{2!} + \ldots + \dfrac{x^n}{n!} + \dfrac{e^c}{(n+1)!} x^{n+1}$

2. $\sin x = x - \dfrac{x^3}{3!} + \dfrac{x^5}{5!} - \ldots + (-1)^n \dfrac{x^{2n-1}}{(2n-1)!} + \dfrac{\sin^{(2n+1)} c}{(2n+1)!} x^{2n+1}$

3. $\cos x = 1 - \dfrac{x^2}{2!} + \dfrac{x^4}{4!} - \ldots + (-1)^n \dfrac{x^{2n}}{(2n)!} + \dfrac{\cos^{(2n+2)} c}{(2n+2)!} x^{2n+2}$

4. $\ln(1+x) = x - \dfrac{x^2}{2} + \dfrac{x^3}{3} - \ldots + \dfrac{(-1)^{n-1}}{n} x^n + \dfrac{x^{n+1}}{(n+1)(1+c)}$,
   $|x| < 1$

4.12 Investigations of Function

a. **Intervals of Increase and Decrease of a Function:**

1. Differentiable Function $f(x)$ is constant in interval (a,b) if, and only if, $f'(x) \equiv 0$ for all $a < x < b$.

2. $f(x)$ is not decreasing in interval (a,b) if, and only if, $f'(x) \geq 0$.

3. $f(x)$ is not increasing in interval (a,b) if, and only if, $f'(x) \leq 0$.

4. $f(x)$ is increasing in interval (a,b) if, and only if, $f'(x) > 0$ for all $x \in (a,b)$.

5. $f(x)$ is decreasing in interval (a,b) if, and only if, $f'(x) < 0$, for all $x \in (a,b)$.

Note: Propositions 4 and 5 are true one-way only. That is, if $f(x)$ is differentiable and monotone increasing in interval (a,b), then, it doesn't necessarily follow that $f'(x) > 0$ for all points of

(a,b). For example, $f(x) = x^3$ is increasing in $(-1,1)$ yet $f'(0) = 0$.

b. **Local Maximum and Minimum Values:**

1. $f(x)$ has **a local minimum value** at x_0 if there exists a definite neighborhood of x_0 where there holds $f(x) \leq f(x_0)$ for all x in this neighborhood.

2. $f(x)$ has **a local maximum value** at x_0 if there exists a definite neighborhood of x_0 where there holds $f(x) \leq f(x_0)$ for all x in this neighborhood.

    Point x_0, which is a local minimum or maximum, is called a local extreme point or **local extremum** of $f(x)$.

3. A necessary condition for the existence of local extremum: If function $f(x)$ is differentiable in then neighborhood of extreme point x_0, then $f'(x_0) = 0$.

4. x_0 is called a **critical point** of $f(x)$ if $f'(x_0) = 0$. Critical points and points in which $f(x)$ is not differentiable are called **suspected extremum points**.

c. **Sufficient Condition of Extremum**

1. Let $f(x)$ be a function defined in the neighborhood of x_0. The **sign of** $f(x)$ **is said to change from negative to positive** at x_0, if there exists a neighborhood $(x_0 - \varepsilon, x_0 + \varepsilon)$, such that for all x in the interval $(x_0 - \varepsilon, x_0)$, $f(x) < 0$, and, for all x in the interval $(x_0, x_0 + \varepsilon)$, $f(x) > 0$. The same way we determine when the sign of $f(x)$ changes from negative to positive at x_0. $f(x)$ **maintains its sign** in point x_0, if there exists a neighborhood $(x_0 - \varepsilon, x_0 + \varepsilon)$

where $f(x) > 0$ for all x of, or $f(x) < 0$ for all x in this neighborhood.

2. **First Derivative Test**

Let x_0 be a suspected extremum point of function $f(x)$, if $f(x)$ is continuous in x_0, and is differentiable in the neighborhood of x_0, except, possibly, in x_0, then follows:

a) If the sign of $f'(x)$ changes from negative to positive in x_0, then x_0 is a **local minimum** of $f(x)$.

b) If the sign of $f'(x)$ changes from positive to negative in x_0, then x_0 is a **local maximum** of $f(x)$.

c) If $f'(x)$ maintains its sign at x_0, then x_0 is **not** a local extremum of $f(x)$.

3. **Second Derivative Test**

If x_0 is a critical point of $f(x)$ and $f(x)$ is twice differentiable, then follows:

a) If $f''(x_0) > 0$ then x_0 is a local minimum of $f(x)$.

b) If $f''(x_0) < 0$ then x_0 is a local maximum of $f(x)$.

c) If $f''(x_0) = 0$, then we cannot conclude on x_0 from this method, and we should examine it using the previous method or other ways.

d. **Absolute Maximum and Minimum in Domain** D

1. Point $x_0 \in D$ is an **absolute maximum** of $f(x)$ in domain D if, for all $x \in D$, there holds $f(x) \le f(x_0)$.

2. Point x_0 is called **absolute minimum** of $f(x)$ in domain D if, for all $x \in D$, there holds $f(x_0) \le f(x)$.

e. **Concavity and Points of Inflection**

1. Function $f(x)$ is **concave down** on interval (a,b), if, for all $x_1, x_2 \in (a,b)$ and for all $0 < t < 1$, there holds

$$f\left(tx_1 + (1-t)x_2\right) \ge tf(x_1) + (1-t)f(x_2)$$

2. Function $f(x)$ is **concave up** on interval (a,b), if, for all $x_1, x_2 \in (a,b)$ and for all $0 < t < 1$, there holds

$$f\left(tx_1 + (1-t)x_2\right) \le tf(x_1) + (1-t)f(x_2)$$

3. If function $f(x)$ is differentiable on x_0, and there exists an neighborhood of x_0 where the graph of the function is under the tangent line to the graph at that point, that is, there exists an $\varepsilon > 0$ such that, for all x which holds $0 < |x - x_0| < \varepsilon$ there holds $f(x) < f'(x_0)(x - x_0) + f(x_0)$, Then $f(x)$ is **concave down** on x_0.

4. Function $f(x)$ is **concave up** on x_0, if there exists a neighborhood of x_0 where the graph of $f(x)$ is above the tangent line to the graph at that point.

5. Function $f(x)$ is **concave down** on open interval (a,b), if it is concave down on any point of the interval. Similarly, we define concavity up on interval (a,b).

6. Sufficient Condition of Concavity

Let $f(x)$ be a function twice differentiable on interval (a,b), then:

a) If, for all x in interval (a,b), $f''(x) > 0$, then $f(x)$ is concave up in interval (a,b).

b) If, for all x in interval (a,b), $f''(x) < 0$, then $f(x)$ is concave down in interval (a,b).

f. Points of Inflection

Point x_0 is the point of inflection of function $f(x)$, if $f(x)$ is continuous at x_0 and there exists a neighborhood $(x_0 - \varepsilon, x_0 + \varepsilon)$ such that the concavity directions of the function in intervals $(x_0 - \varepsilon, x_0)$, $(x_0, x_0 + \varepsilon)$ are opposite. That is, if $f(x)$ is concave down in interval $(x_0 - \varepsilon, x_0)$ and concave up in interval $(x_0, x_0 + \varepsilon)$, or vise versa.

g. Asymptotes

1. The straight line $x = a$ is a **vertical asymptote** of function $f(x)$, which is defined in a right-handed or left-handed neighborhood of point $x = a$, except, possibly, a, if, at least one of the limits $\lim\limits_{x \to a^+} f(x)$, $\lim\limits_{x \to a^-} f(x)$ equals to ∞ or $-\infty$.

2. The geometric meaning of the existence of vertical asymptote is that the graph of $f(x)$, near point $x = a$, gets steep and very close to the straight line $x = a$, but doesn't contact it.

3. The straight line $y = ax + b$ is an **oblique asymptote** of $f(x)$ at $+\infty$, when $\lim\limits_{x \to \infty}[f(x) - (ax + b)] = 0$.

If $a = 0$, the asymptote is also called a **horizontal asymptote** of $f(x)$, since $y = b$ is a horizontal straight line.

4. The straight line $y = mx + n$ is an **oblique asymptote** at $-\infty$, if

$$\lim_{x \to -\infty} [f(x) - (mx + n)] = 0$$

5. If $f(x)$ is a function defined in interval (α, ∞), and if the limits $a = \lim_{x \to \infty} \dfrac{f(x)}{x}$, $b = \lim_{x \to \infty}[f(x) - ax]$ exist, then the straight line $y = ax + b$ is the unique oblique asymptote of $f(x)$ when $x \to \infty$.

6. If $f(x)$ is a function defined in interval $(-\infty, c)$, and if the limits $m = \lim_{x \to -\infty} \dfrac{f(x)}{x}$, $n = \lim_{x \to -\infty}[f(x) - mx]$, then the straight line $y = mx + n$ is the unique oblique asymptote of $f(x)$ in $-\infty$.

h. **Investigation of a Function**

Main stages in investigation of a function:

1. The domain of the function
2. The intersection points of the graph with the coordinate axes
3. Extreme
4. Intervals of increase and decrease
5. Intervals of concavity
6. Points of inflection
7. Asymptotes
8. Gathering the data in a table
9. Drawing then graph of the function

V. Integral Calculus of Single-Variable Functions

Chapter 1
Indefinite Integral

1.1   Antiderivative

a. The function $F(x)$ is the **antiderivative** of $f(x)$ in domain D, if, for all x of D, there holds $F'(x) = f(x)$.

b. If $F(x)$ is an antiderivative of $f(x)$, then, for all constant c, $F(x) + c$ is also an antiderivative of $f(x)$.

c. The set of all antiderivatives $F(x) + c$ of $f(x)$ is called the **indefinite integral** of $f(x)$. It is denoted

$$\int f(x) dx = F(x) + c$$

1.2   Properties of Integral

a. $\int f'(x) dx = f(x) + c$

b. $\int a f(x) dx = a \int f(x) dx$

c. $\int [f(x) \pm g(x)] dx = \int f(x) dx \pm \int g(x) dx$

1.3   Immediate Integrals Table

a. $\int x^{\alpha}dx = \dfrac{x^{\alpha+1}}{\alpha+1} + c, \ \alpha \neq -1$ b. $\int \dfrac{dx}{x} = \ln|x| + c$

c. $\int \sin x dx = -\cos x + c$ d. $\int \cos x dx = \sin x + c$

e. $\int \dfrac{dx}{\cos^2 x} = \tan + c$ f. $\int \dfrac{dx}{\sin^2 x} = -\cot x + c$

g. $\int a^x dx = \dfrac{a^x}{\ln a} + c$ h. $\int e^x dx = e^x + c$

i. $\int \dfrac{dx}{x^2 + a^2} = \dfrac{1}{a}\arctan\dfrac{x}{a} + c$ j. $\int \dfrac{dx}{\sqrt{a^2 - x^2}} = \arcsin\dfrac{x}{a} + c$

k. $\int \sin hx dx = \cos hx + c$ l. $\int \cos hx dx = \sin hx + c$

1.4   Integration by Parts

$$\int u(x) \cdot v'(x) dx = u(x) \cdot v(x) - \int u'(x) \cdot v(x) dx$$

Example: To calculate $\int x \sin x dx$, let's write $u(x) = x$, and $v'(x) = \sin x$.

Therefore, $u'(x) = 1$ and $v(x) = -\cos x$.

Substituting in the formula, we get

$$\int x \sin x dx = -x \cdot \cos x - \int (-\cos x) \cdot (x)' dx = -x \cos x + \int \cos x dx =$$

$$= -x \cos x + \sin x + c$$

1.5   Substitution Method

a. This way, we substitute variable x in the function with a function of another variable, t: $x = \varphi(t)$. This way, the integral in the right side is simpler.

$$\int f(x)dx = \int f(\varphi(t)) \cdot \varphi'(t)dt$$

Finding out the integral, we return to variable x. The equality is meaningful if function $\varphi(t)$ is invertible.

Example: To calculate $\int \dfrac{dx}{(2-3x)^2}$, let's substitute

$t = 2-3x$. Therefore, $dx = -\dfrac{1}{3}dt$, and the result is

$$\int \frac{dx}{(2-3x)^2} = -\frac{1}{3}\int \frac{dt}{t^2} = \frac{1}{3t} + c = \frac{1}{3(2-3x)} + c$$

b. For integrals in the form of $\int x^n \cdot \sqrt[m]{ax+b}\,dx$, $m,n \in \mathbb{Z}$, the substitution is $t^m = ax+b$.

c. For integrals in the form of $\int x^{2n+1} \cdot \sqrt{a^2 \pm x^2}\,dx$, $n \in \mathbb{Z}$, the substitution is $t^2 = a^2 \pm x^2$.

d. For integrals in the form of $\int R\left(x, \sqrt[m]{\dfrac{ax+b}{cx+d}}\right)dx$, $m \in \mathbb{Z}$, when R is a rational function of two-variables, the substitution is $\dfrac{ax+b}{cx+d} = t^m$, from there follows $x = \dfrac{dt^m - b}{a - ct^m}$.
After substitution in the integral, we get a new integral containing no roots.

e. For integrals in the form of $\int x^{2n}(a^2 - x^2)^{\pm\frac{1}{2}}dx$, the substitution is $x = a\sin t$ or $x = a\cos t$.

f. For integrals in the form of $\int x^{2n}(a^2 + x^2)^{\pm\frac{1}{2}}dx$, the substitution is $x = a\tan t$, $x = a\sinh t$

g. For integrals in the form of $\int x^{2n}(x^2-a^2)^{\pm\frac{1}{2}}dx$, the substitution is $x=\dfrac{a}{\sin t}$, $x=a\cosh t$.

1.6   Trigonometric Functions Integration

Calculating the integrals of trigonometric function involves using the relevant trigonometric identities.

a. **Integrals**

$$\int \sin\alpha x \cos\beta x \, dx \, , \, \int \sin\alpha x \sin\beta x \, dx \, , \, \int \cos\alpha x \cos\beta x \, dx$$

In each of these cases, we should use the relevant trigonometric identities (see II, 2.6 and Chapters 8-10), turning the multiplication into summation.

b. For integral in the form of $\int \sin^{2m+1}x\cos^n x\, dx$, $m,n\in\mathbb{Z}$, the substitution is $t=\cos x$.

c. For integral in the form of $\int \sin^m x\cos^{2n+1}x\, dx$, $m,n\in\mathbb{Z}$, the substitution is $t=\sin x$.

d. For integral in the form of $\int \sin^m x\cos^n x\, dx$, when $m,n\in\mathbb{N}$ one or both are even and the other is zero, we use the formulas $2\cos^2 x=1+\cos 2x$, $2\sin^2 x=1-\cos 2x$.

e. For integral in the form of $\int R(\sin x,\cos x)dx$, when R is a rational function, the substitution is $t=\mathrm{tg}\dfrac{x}{2}$.

In this case, $\cos x=\dfrac{1-t^2}{1+t^2}$, $\sin x=\dfrac{2t}{1+t^2}$, $dx=\dfrac{2dt}{1+t^2}$.

Example: To calculate $\int\dfrac{dx}{\cos x+2\sin x+3}$, we use the substitution $t=\tan\dfrac{x}{2}$.

$$\int \frac{dx}{\cos x + 2\sin x + 3} = \int \frac{\dfrac{2dt}{1+t^2}}{\dfrac{1-t^2}{1+t^2} + \dfrac{4t}{1+t^2} + 3} = \int \frac{2dt}{1-t^2 + 4t + 3 + 3t^2} =$$

$$= \int \frac{2dt}{2t^2 + 4t + 4} = \int \frac{dt}{t^2 + 2t + 2} = \int \frac{dt}{(t+1)^2 + 1} =$$

$$= \arctan(t+1) = \arctan\left(1 + \tan\frac{x}{2}\right) + c$$

1.7   Rational Functions Integration

a. Basic Rational Functions of Two Types

1. $\dfrac{A}{(x-a)^n}$, $n \in \mathbb{N}$

2. $\dfrac{Ax+B}{(x^2+px+q)^n}$, $\Delta = p^2 - 4q < 0$, $n \in \mathbb{N}$

b. Integration of basic rational function of the first type:

1. $\int \dfrac{Adx}{x-a} = A \ln|x-a| + c$

2. $\int \dfrac{A}{(x-a)^n} = \dfrac{A}{(1-n)}(x-a)^{1-n} + c$, $n > 1$

c. Integration of basic rational function of the second type:

If $p^2 - 4q < 0$, then, after completing the square to

$x^2 + px + q = \left(x + \dfrac{p}{2}\right)^2 + q - \dfrac{p^2}{4}$ and the substitution of

$a^2 = q - \dfrac{p^2}{4}$, $t = x + \dfrac{p}{2}$, we get

$$\int \frac{Ax+B}{x^2+px+q}dx = A\int \frac{tdt}{t^2+a^2} + \frac{2B-Ap}{2}\int \frac{dt}{t^2+a^2} =$$

$$= \frac{A}{2}\ln(x^2+px+q) - \frac{2(2B-Ap)}{p^2-4q}\arctan\frac{2x+p}{\sqrt{4q-p^2}}+c$$

d. For a recursive formula for finding $\int \dfrac{Ax+B}{(x^2+px+q)^n}dx$,

$n>1$, see integrals table in (XVI, 1.4).

e. **A General Case**

If, in rational function $\dfrac{P(x)}{Q(x)}$, the degree of polynomial

$P(x)$ is higher than or equal to that of polynomial $Q(x)$, that is, the function is not common, we divide $P(x)$ by $Q(x)$, and the result is the sum of polynomial $M(x)$ and a common rational function (see X. 3.3)

$$\frac{P(x)}{Q(x)} = M(x) + \frac{R(x)}{Q(x)}$$

Step 1: Break down the polynomial $Q(x)$ into factors of the form of $(x-\alpha_i)^{m_i}$ and $(x^2+p_jx+q_j)^{n_j}$, when $p_j^2-4q_j<0$ (that is, no real roots exist):

$$Q(x)=(x-\alpha_1)^{m_1}\ldots(x-\alpha_k)^{m_k}(x^2+p_1x+q_1)^{n_1}\ldots(x^2+p_\ell x+q_\ell)^{n_\ell}$$

Step 2: Break down the common fraction $\dfrac{R(x)}{Q(x)}$ into a sum of unit fractions, the following way:

$$\frac{R(x)}{Q(x)} = \left[\frac{A_1^1}{(x-\alpha_1)} + \frac{A_2^1}{(x-\alpha_1)^2} + \ldots + \frac{A_{m_1}^1}{(x-\alpha_1)^{m_1}}\right] +$$

$$+\left[\frac{A_1^2}{(x-\alpha_2)} + \frac{A_2^2}{(x-\alpha_2)^2} + \ldots + \frac{A_{m_2}^2}{(x-\alpha_2)^{m_2}}\right] +$$

$$+...+\left[\frac{A_1^k}{(x-\alpha_k)}+\frac{A_2^k}{(x-\alpha_k)^2}+...+\frac{A_{mk}^k}{(x-\alpha_k)^{mk}}\right]+$$

$$+\left[\frac{B_1^1x+C_1^1}{x^2+p_1x+q_1}+\frac{B_2^1x+C_2^1}{(x^2+p_1+q_1)^2}+...+\frac{B_{n_1}^1x+C_{n_1}^1}{(x^2+p_1x+q_1)^{n_1}}\right]+$$

$$+...+\left[\frac{B_1^\ell x+C_1^\ell}{x^2+p_\ell x+q_\ell}+\frac{B_2^\ell x+C_2^\ell}{(x^2+p_\ell+q_\ell)^2}+...+\frac{B_{n_\ell}^\ell x+C_{n_\ell}^\ell}{(x^2+p_\ell x+q_\ell)^{n_\ell}}\right]$$

The computing of quotients A_j^i, B_j^i, C_j^i, is done by adding up all the fractions on the right side into one fraction, the denominator of which is $Q(x)$. Then, by equating the denominator of this fraction to the quotients of $R(x)$, we get a system of linear equations, with the unknowns A_j^i, B_j^i, C_j^i.

Step 3: Find the integral of the rational function this way:

$$\int\frac{P(x)}{Q(x)}dx=\int M(x)dx+\int\frac{R(x)}{Q(x)}dx$$

In the last integral, substitute $\frac{R(x)}{Q(x)}$ with the right side of the equality of Step 2. This way, we get the sum of unit fractions integrals.

Example: To find the integral $\int\frac{5x^2+6x+9}{(x-3)^2(x+1)^2}dx$, we break down the rational function into unit fractions:

$$\frac{5x^2+6x+9}{(x-3)^2(x+1)^2}=\frac{A}{x-3}+\frac{B}{(x-3)^2}+\frac{C}{x+1}+\frac{D}{(x+1)^2}$$

By adding the fractions of the right side, and equating numerators, we get:

$$5x^2+6x+9=A(x-3)(x+1)^2+B(x+1)^2+$$

$$+C(x+1)(x-3)+D(x-3)^2$$

After equating quotients, the result is

$$A = C = 0, \ B = \frac{1}{2}, \ D = \frac{9}{2}$$

$$\int \frac{5x^2 + 6x + 9}{(x-3)^2(x+1)^2} \, dx = \int \left(\frac{9}{2(x-3)^2} + \frac{1}{2(x+1)^2} \right) dx =$$

$$= -\frac{9}{2(x-3)} - \frac{1}{2(x+1)} + c$$

Chapter 2
Definite Integral

2.1   Definition

Let $f(x)$ be a function defined on interval $[a,b]$. Let's divide interval $[a,b]$ into subintervals by the points

$$a = x_0 < x_1 < x_2 < \ldots < x_{n-1} < x_n = b$$

We denote $\Delta x_i = [x_{i-1}, x_i]$, using the same denotation for the length of all subintervals, $\Delta x_i = x_i - x_{i-1}$, $i = 1, 2, \ldots, n$.

Let us denote $\Delta(T) = \max(\Delta x_1, \Delta x_2, \ldots, \Delta x_n)$. In every Δx_i, we select an arbitrary point c_i. The expression

$$\sum_{i=1}^{n} f(c_i) \Delta x_i = f(c_1) \Delta x_1 + f(c_2) \Delta x_2 + \ldots + f(c_n) \Delta x_n$$

is called **Riemann sum**, according to the partition to subintervals and the selected points c_i.

Definition: function $f(x)$ is called a **Riemann integrable function**, on interval $[a,b]$, if the limit $I = \lim\limits_{\Delta(T)\to 0} \sum\limits_{i=1}^{n} f(c_i) \cdot \Delta x_i$ exists and is not dependent on the selection of partitions and points c_i. We denote this limit by $I = \int\limits_{a}^{b} f(x)\,dx$, and it is called the **definite integral** of function $f(x)$ on interval $[a,b]$, when a and b in integral I are called **limits of integration**.

2.2   Classes of Integrable Functions

a. If function $f(x)$ is not bounded on interval $[a,b]$, then it is not integrable on that interval.

b. If function $f(x)$ is continuous in interval $[a,b]$, then it is integrable in it.

c. A function defined and bounded in interval $[a,b]$ is called **piecewise continuous** function, if it has at most a finite set of discontinuities, all of which are jump discontinuities.

d. If function $f(x)$ is piecewise continuous in interval $[a,b]$, then it is integrable in it.

2.3   Properties of Definite Integral

a. $\int\limits_{b}^{a} f(x)\,dx = -\int\limits_{a}^{b} f(x)\,dx$

b. $\int\limits_{a}^{b} [f(x) \pm g(x)]\,dx = \int\limits_{a}^{b} f(x)\,dx \pm \int\limits_{a}^{b} g(x)\,dx$

c. $\int\limits_{a}^{b} cf(x)\,dx = c\int\limits_{a}^{b} f(x)\,dx$ for every constant c

d. $\int\limits_{a}^{c} f(x)\,dx = \int\limits_{a}^{b} f(x)\,dx + \int\limits_{b}^{c} f(x)\,dx$

e. If $m \le f(x) \le M$, $x \in [a,b]$, then

$m(b-a) \le \int\limits_{a}^{b} f(x)\,dx \le M(b-a)$, particularly when

$M = \sup\limits_{x \in [a,b]} f(x)$, $m = \inf\limits_{x \in [a,b]} f(x)$.

f. If $f(x)$ and $g(x)$ are integrable on $[a,b]$, then $f(x)g(x)$ is integrable on interval $[a,b]$.

g. **Mean Value Theorem for Integrals**

If $f(x)$ is continuous on $[a,b]$, then there is a point c on $[a,b]$, such that

$$\int\limits_{a}^{b} f(x)\,dx = f(c)(b-a)$$

The number $f(c)$ is called **the average value of $f(x)$ on** $[a,b]$.

h. If $f(x) \ge g(x)$ for all $x \in [a,b]$, then $\int\limits_{a}^{b} f(x)\,dx \ge \int\limits_{a}^{b} g(x)\,dx$.

i.

1. If $f(x)$ is integrable on $[a,b]$, then $|f(x)|$ is also integrable in that interval and there holds

$\left| \int\limits_{a}^{b} f(x)\,dx \right| \le \int\limits_{a}^{b} |f(x)|\,dx$.

2. The inverse is incorrect. That is, if $|f(x)|$ is integrable, it doesn't necessarily follow that $f(x)$ is integrable in that interval.

j. If $g(x) \ge 0$ and $m \le f(x) \le M$ for all $x \in [a,b]$, then

$$m\int_a^b g(x)\,dx \le \int_a^b f(x)g(x)\,dx \le M\int_a^b g(x)\,dx$$

2.4   Connection Between the Indefinite and Definite Integral

If $f(x)$ is a continuous function and $F(x)$ is its antiderivative on $[a,b]$, then:

a. $F(x) = \int_c^x f(t)\,dt$, when $[c,x] \subset [a,b]$

b. $\dfrac{d}{dx}\int_c^x f(t)\,dt = f(x)$

c.   **Newton-Leibniz Formula** $\int_a^b f(x)\,dx = F(x)\big|_a^b = F(b) - F(a)$

d. $\dfrac{d}{dx}\int_{\alpha(x)}^{\beta(x)} f(t)\,dt = f\big(\beta(x)\big)\cdot\beta'(x) - f\big(\alpha(x)\big)\cdot\alpha'(x)$

e.   **Cauchy-Schwartz Inequality**

   If $f(x)$ and $g(x)$ are integrable on $[a,b]$, then:

$$\int_a^b |f(x)\cdot g(x)|\,dx \le \sqrt{\int_a^b |f(x)|^2\,dx} \cdot \sqrt{\int_a^b |g(x)|^2\,dx}$$

2.5   Calculating Definite Integrals

a.   Change of Variables

If function $f(x)$ is continuous on $[a,b]$, and if function $x = g(t)$ is continuously differentiable on $[\alpha,\beta]$, if its image equals interval $[a,b]$, and there holds $g(\alpha) = a$ and $g(\beta) = b$, then:

$$\int_a^b f(x)\,dx = \int_\alpha^\beta f\big(g(t)\big)\cdot g'(t)\,dt$$

b. **Integration by Parts**

If functions $u(x)$ and $v(x)$ have continuous derivatives on $[a,b]$, then:

$$\int_a^b u(x)v'(x)\,dx = u(x)v(x)\big|_a^b - \int_a^b u'(x)v(x)\,dx$$

c. **Integral of Even and Odd Functions on Interval $[-a,a]$**

If function $f(x)$ **is even and integrable** on $[-a,a]$, then:

$$\int_{-a}^a f(x)\,dx = 2\int_0^a f(x)\,dx$$

And, if function $f(x)$ **is odd and integrable** on $[-a,a]$, then:

$$\int_{-a}^a f(x)\,dx = 0$$

2.6   Numerical Methods of Computing Definite Integrals

We divide interval $[a,b]$ into n equal sub-intervals, by points:

$$x_0 = a\ ,\ x_1 = a+h, x_2 = a+2h,\ldots,x_n = a+nh = b$$

when $h = \dfrac{b-a}{n}$, $\Delta x_i = x_i - x_{i-1}$.

a. The following formulas are **rectangle approximations**:

$$I = \int_a^b f(x)\,dx \approx \frac{b-a}{n}\big[f(x_0)+f(x_1)+f(x_2)+\ldots+f(x_{n-1})\big] = I_n$$

$$I = \int_a^b f(x)\,dx \approx \frac{b-a}{n}\left[f(x_1)+f(x_2)+f(x_3)+...+f(x_n)\right] = I_n$$

$$I = \int_a^b f(x)\,dx \approx \frac{b-a}{n}\left[f(c_1)+f(c_2)+f(c_3)+...+f(c_n)\right] = I_n$$

when c_i is the midpoint of $[x_{i-1}, x_i]$.

The approximation error, in the case that $f(x)$ is twice differentiable is $|I - I_n| \leq \dfrac{(b-a)^3}{24n^2}\max\limits_{a \leq x \leq b}|f''(x)|$.

b. **Trapezoid Approximation**

$$I = \int_a^b f(x)\,dx \approx \frac{b-a}{n}\left[\frac{f(x_0)+f(x_n)}{2}+f(x_1)+f(x_2)+...+f(x_{n-1})\right] = I_n$$

The **approximation error** is $|I - I_n| \leq \dfrac{(b-a)^3}{12n^2}\max\limits_{a \leq x \leq b}|f''(x)|$

c. **Simpson's Rule**

$$I = \int_a^b f(x)\,dx \approx \frac{b-a}{3n}\left[\frac{f(x_0)+f(x_n)}{2}+f(x_1)+f(x_2)+...\right.$$

$$\left. +f(x_n)+2[f(c_1)+f(c_2)+...+f(c_n)]\right] = I_n$$

when c_i is then midpoint of $[x_{i-1}, x_i]$.

The approximation error is

$$|I - I_n| \leq \frac{(b-a)^5}{180 \cdot (2n)^4}\max\limits_{a \leq x \leq b}|f^{(4)}(x)|$$

2.7   Applications of Definite Integral

a. The **area of a plane confined** by the graph of non-negative function $f(x)$, x-axis, and straight lines $x = a$ and $x = b$ is

$$S = \int_a^b f(x)\,dx$$

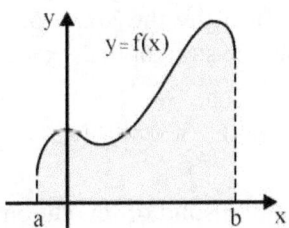

b. The **Area of a plane bounded** by a curve presented in the parametric form $x = \varphi(t), y = \psi(t), t \in [\alpha, \beta]$ is

$$S = \int_\alpha^\beta \psi(t) \cdot \varphi'(t)\,dt$$

c. The area of a plane bounded by two functions $f(x)$, $g(x)$, above interval $[a,b]$ is

$$S = \int_a^b |f(x) - g(x)|\,dx$$

d. The area of a plane bounded by a curve presented by $\rho = f(\theta)$, $\alpha \le \theta \le \beta$, and two rays, $\theta = \alpha$ and $\theta = \beta$ is

$$S = \frac{1}{2}\int_\alpha^\beta |f(\theta)|^2\,d\theta$$

e. The length of a planar curve given by:

1) $y = f(x)$, $x \in [a,b]$ is $L = \int_a^b \sqrt{1 + [f'(x)]^2}\,dx$

2) $x = \varphi(t), y = \psi(t)$, $t \in [\alpha, \beta]$ is $L = \int\limits_{\alpha}^{\beta} \sqrt{\varphi'^2(t) + \psi'^2(t)}\, dx$

f. The volume of a solid of revolution around the x-axis of an area confined by the graph of non-negative function $y = f(x)$ and the straight lines $x = a$, $x = b$, is

$$V = \pi \int\limits_{a}^{b} [f(x)]^2\, dx$$

g. The volume of a solid of revolution around the x-axis of an area confined by the graphs of functions $f(x)$ and $g(x)$ and straight lines $x = a$, $x = b$, is

$$V = \pi \int\limits_{a}^{b} [f(x) - g(x)]^2\, dx$$

h. The volume of a solid of revolution around the y-axis of an area confined by the graph of non-negative function $y = f(x)$ and the straight lines $x = a$, $x = b$, is

$$V = 2\pi \int\limits_{a}^{b} x f(x)\, dx$$

i. The volume of a solid of revolution around the y-axis of an area confined by the graphs of functions $f(x)$ and $g(x)$ in interval $[a, b]$ is

$$V = 2\pi \int\limits_{a}^{b} x \,|\, g(x) - f(x) \,|\, dx$$

j. If a solid body is confined by planes $x = a$, $x = b$, and the plane $x = x_0$ carves out of the body an area $s(x_0)$ for every $x_0 \in [a, b]$, then, the volume of the body is

$$V = \int\limits_{a}^{b} s(x)\, dx$$

Example: Calculate the volume of the sphere $x^2 + y^2 + z^2 = R^2$, confined between the planes $x = R$ and $x = -R$.

For every x situated at $[-R, R]$, the section is circle $y^2 + z^2 = R^2 - x^2$, the area of which is $s(x) = \pi(R^2 - x^2)$.
Therefore, the volume of the sphere is

$$V = \pi \int_{-R}^{R} (R^2 - x^2)\,dx = \pi \left(R^2 x - \frac{x^3}{3} \right)_{-R}^{R} = \frac{4\pi}{3} R^3$$

Chapter 3
Improper Integral

Definite Integral is defined on finite interval $[a, b]$, and function $f(x)$ must be bounded in it.

An **improper integral** is defined on an infinite interval such as $[a, \infty)$, $(-\infty, b]$, as well as on the entire straight line $(\infty, -\infty)$.

In addition, it is also defined on not necessarily bounded functions.

3.1   Integrals with Infinite Limits

a. The improper integral $\int_{a}^{\infty} f(x)\,dx$ of function $f(x)$ defined on $[a, \infty)$, and integrable on $[a, b]$ for all $b > a$, **exists** or **converges**, if the limit $\lim\limits_{b \to \infty} \int_{a}^{b} f(x)\,dx$ exists and is finite, otherwise, we say the integral is **not-convergent or divergent**.

We write $\int\limits_a^\infty f(x)\,dx = \lim\limits_{b\to\infty}\int\limits_a^b f(x)\,dx$

b. The improper integral on $(-\infty, a]$ is

$$\int\limits_{-\infty}^a f(x)\,dx = \lim\limits_{b\to-\infty}\int\limits_b^a f(x)\,dx$$

c. Let $f(x)$ be integrable in all closed interval $[-b, b]$, at any point $a \in (-\infty, \infty)$. Let us define

$$\int\limits_{-\infty}^\infty f(x)\,dx = \int\limits_{-\infty}^a f(x)\,dx + \int\limits_a^\infty f(x)\,dx$$

If both of the integrals on the right are convergent, then we say the integrals on the left side are convergent. If at least one of the integrals on the right side is divergent, then the integral on the left is divergent. If the integral on the left side is divergent, then it is not dependent on the selection of a .

d. **Convergence Tests**

1. Let $f(x)$ be integrable on $[a, b]$ for all $b > a$, when a is a constant. Let $M > 0$ and α be two numbers.

If $\alpha > 1$ and $0 \le f(x) \le \dfrac{M}{x^\alpha}$ for all x in $[a, \infty)$, then the integral $\int\limits_a^\infty f(x)\,dx$ is convergent.

If $\alpha \le 1$ and $\dfrac{M}{x^\alpha} \le f(x)$ for all x in $[a, \infty)$, then the integral $\int\limits_a^\infty f(x)\,dx$ is divergent.

2. Cauchy Criterion for Convergence of an Improper Integral

Let function $f(x)$ be integrable on $[a,b]$ for all $b > a$. Then, integral $\int_a^\infty f(x)\,dx$ is convergent if, and only if, for every $\varepsilon > 0$ there exist b_0 such that for all $b_1, b_2 \geq b_0$, there holds $\left| \int_{b_1}^{b_2} f(x)\,dx \right| < \varepsilon$.

A similar proposition holds for integral $\int_{-\infty}^b f(x)\,dx$.

3. Comparison Test

Let $f(x)$ and $g(x)$ be two non-negative functions defined on $[a,\infty)$, and integral on $[a,b]$ for every $b > a$. If, for all $x \geq b_0$, there holds $f(x) \leq g(x)$, then, from the convergence of integral $\int_a^\infty g(x)\,dx$, there follows the convergence of integral $\int_a^\infty f(x)\,dx$.

If integral $\int_a^\infty f(x)\,dx$ is divergent, then $\int_a^\infty g(x)\,dx$, is also divergent.

4. Ratio Test

Let $f(x)$ and $g(x)$ be non-negative functions on $[a,\infty)$. If the limit $L = \lim\limits_{x \to \infty} \dfrac{f(x)}{g(x)}$ exists, then when:

a) $0 < L < \infty$, the integrals $\int_a^\infty f(x)\,dx$ and $\int_a^\infty g(x)\,dx$ converge or diverge together. That is, of one integral is convergent, then the other is convergent, and if one of them is divergent, than the other is divergent.

b) $L = 0$, then from the convergence of integral $\int_a^\infty g(x)dx$, there follows the convergence of integral $\int_a^\infty f(x)dx$.

c) $L = \infty$, then from the convergence of integral $\int_a^\infty f(x)dx$, there follows the convergence of integral $\int_a^\infty g(x)dx$.

5. **Abel Test**

Let $f(x)$ and $g(x)$ be functions integrable on $[a,b]$, for all $b > a$. If the integral $\int_a^\infty f(x)dx$ is convergent, and if $g(x)$ is monotone and bounded on interval $[a, \infty]$, then the integral $\int_a^\infty f(x)g(x)dx$ is convergent.

6. **Dirichlet Test**

Let $f(x)$ and $g(x)$ be functions integrable on $[a,b]$, for all $b > a$. Suppose there exists an M such that $\left| \int_a^b f(x)dx \right| \le M$ for all b, if $g(x)$ is monotone on $[a, \infty)$ and $\lim\limits_{x \to \infty} g(x) = 0$. Then, the integral $\int_a^\infty f(x)g(x)dx$ is convergent.

e. Absolute Integrability

1. Function $f(x)$ is **absolutely integrable** on $[a,\infty)$, if the improper integral $\int\limits_{a}^{\infty} |f(x)|\, dx$ is convergent.

    We also say that integral $\int\limits_{a}^{\infty} f(x)\, dx$ is **absolutely convergent.**

2. If $f(x)$ is absolutely integrable on interval $[a,\infty)$, then it is integrable on that interval. That is, if the integral $\int\limits_{a}^{\infty} |f(x)|\, dx$ is convergent, then integral $\int\limits_{a}^{\infty} f(x)\, dx$ is also convergent.

3.2   Unbound Function Integral

a. Let $f(x)$ defined and unbounded on semi-open interval $[a,b)$, and, for all $\varepsilon > 0$, the integral $\int\limits_{a}^{b-\varepsilon} f(x)\, dx$ exists. If the limit $I = \lim\limits_{\varepsilon \to 0^+} \int\limits_{a}^{b-\varepsilon} f(x)\, dx$ exists, then I is called the improper integral of $f(x)$ and is denoted as $I = \int\limits_{a}^{b} f(x)\, dx$.

b. If $f(x)$ is defined and unbounded on $(a,b]$, and integrable on $[a+\delta, b]$ for all $\delta > 0$, let's define

$$\int\limits_{a}^{b} f(x)\, dx = \lim\limits_{\delta \to 0^+} \int\limits_{a+\delta}^{b} f(x)\, dx$$

c. Let $f(x)$ be:

1. Defined on interval $[a,b]$, except, possibly, at $c \in (a,b)$.

2. Integrable on intervals $(c+\delta,b]$ and $[a,c-\varepsilon)$ for all $\varepsilon, \delta > 0$.

3. Unbounded at the neighborhood of c.

If the limits $I_1 = \lim\limits_{\varepsilon \to 0^+} \int\limits_{a}^{c-\varepsilon} f(x)dx$, $I_2 = \lim\limits_{\delta \to 0^+} \int\limits_{c+\delta}^{b} f(x)dx$ exist, then we say the improper integral of $f(x)$ converges on $[a,b]$, and its value is $\int\limits_{a}^{b} f(x)dx = I_1 + I_2$.

In case $\varepsilon = \delta$, the integral is called **Cauchy principal value integral**, and is denoted $\text{v.p.}\int\limits_{a}^{b} f(x)dx$.

d. **Convergence Test**

1. Let $f(x)$ be a functions defined on interval $[a,b)$ and integrable on $[a,b-\varepsilon]$ for all $\varepsilon > 0$, and let $M > 0$ and α be two real numbers. Then,

a) If $0 \le f(x) \le \dfrac{M}{(b-x)^\alpha}$ for all x on interval $[a,b)$, and if $\alpha < 1$, then integral $\int\limits_{a}^{b} f(x)dx$ is convergent.

b) If $\dfrac{M}{(b-x)^\alpha} \le f(x)$, for all x on interval $[a,b)$, and if $\alpha \ge 1$, then integral $\int\limits_{a}^{b} f(x)dx$ is divergent.

2. If the improper integral $\int_a^b |f(x)| \, dt$ exists, then it is said that $f(x)$ is **absolutely integrable** on $[a,b]$.

3. If $f(x)$ is absolutely integrable on $[a,b]$, then integral $\int_a^b f(x) \, dx$ exists.

3.3   Gamma and Beta Functions

a. Function $\Gamma(p) = \int_0^\infty e^{-u} u^{p-1} \, du$, $p > 0$ is called **gamma function.**

b. Gamma function properties:

1. $\Gamma(p+1) = p\Gamma(p)$

2. $\Gamma(n) = n!$, $n \in \mathbb{N}$, $\Gamma(1) = 1$

3. $\Gamma(\frac{1}{2}) = \sqrt{\pi}$, $\Gamma(p) \cdot \Gamma(1-p) = \dfrac{\pi}{\sin p\pi}$

4. $\Gamma(p)$ is continuous and has continuous differentials at $p > 0$.

5. $\lim\limits_{p \to \infty} \Gamma(p) = \infty$

c. Function $B(a,b) = \int_0^1 x^{a-1}(1-x)^{b-1} \, dx$, $a, b > 0$ is called **beta function.**

d. Beta function properties:

1. $B(a,b) = B(b,a)$

2. $B(a,b) = \dfrac{b-1}{a+b-1} B(a,b-1)$

3. $B(a,n) = \dfrac{n-1}{a+n-1} \cdot \dfrac{n-2}{a+n-2} \cdot \ldots \cdot \dfrac{1}{a+1} \cdot B(a,1)$, $n \in \mathbb{N}$

4. $B(a,1) = \int\limits_{0}^{1} x^{a-1} dx = \dfrac{1}{a}$

5. $B(m,n) = \dfrac{(n-1)!(m-1)!}{(m+n-1)!}$, $m,n \in \mathbb{N}$

6. $B(\frac{1}{2}, \frac{1}{2}) = \pi$

e. The relation between the beta and gamma function

$$B(a,b) = \frac{\Gamma(a) \cdot \Gamma(b)}{\Gamma(a+b)}$$

VI. Series

Chapter 1
Basic Concepts

1.1 Series of Numbers. Summation of Series

a. Let $\{a_n\}_{n=1}^{\infty} = a_1, a_2, \ldots a_n, \ldots$ be a sequence of numbers.

The expression $\sum_{k=1}^{\infty} a_k = a_1 + a_2 + a_3 + \ldots + a_k + \ldots$ is called **infinite series** or just a series, and a_k is the **general term** of the series.

The terms of the series add up to **partial sums**:

$$S_1 = a_1$$
$$S_2 = a_1 + a_2$$
$$S_3 = a_1 + a_2 + a_3$$
$$\vdots$$
$$S_n = a_1 + a_2 + a_3 + \ldots + a_n$$
$$\vdots$$

The result is a sequence of partial sums $\{S_n\}$.

b. Series $\sum_{k=1}^{\infty} a_k$ is **convergent** if there exists a finite limit S

of the sequence of partial sums $\{S_n\}$. S is called the sum

of the series. Written $S = \sum_{k=1}^{\yen} a_k$.

If the limit of $\{S_n\}$ does not exist, or is infinity, the series

is called **divergent**.

c. **Examples**:

1. **Geometric Series** $\sum_{k=0}^{\infty} q^k = 1 + q + q^2 + \ldots + q^n + \ldots$

    converges if and only if $|q| < 1$ and its sum is

    $S = \dfrac{1}{1-q}$.

2. **Leibniz Series** $\ln 2 = \sum_{k=1}^{\infty} \dfrac{(-1)^{k+1}}{k} = 1 - \dfrac{1}{2} + \dfrac{1}{3} - \dfrac{1}{4} + \ldots$

3. Series $\sum_{n=1}^{\infty} (-1)^{n-1} = 1 - 1 + 1 - 1 + \ldots$ is divergent.

4. **Harmonic Series** $\sum_{k=1}^{\infty} \dfrac{1}{k} = 1 + \dfrac{1}{2} + \dfrac{1}{3} + \ldots .$

1.2   Series Remainder

The series $R_n = \sum_{k=n+1}^{\infty} a_k = a_{n+1} + a_{n+2} + \ldots$ is the n-th

remainder of the series $S = \sum_{n=1}^{\infty} a_n$, $S = S_n + R_n$.

1.3   Telescoping Series

a. Let $\{a_n\}_{n=1}^\infty$ be a sequence of numbers. Series $\sum_{n=1}^\infty (a_n - a_{n+1}) = (a_1 - a_2) + (a_2 - a_3) + \ldots$ is called **telescoping series**.

b. A telescoping series is convergent if and only if sequence $\{a_n\}_{n=1}^\infty$ is convergent.

c. If sequence $\{a_n\}_{n=1}^\infty$ converges to a, then the sum of the telescoping series is $S = a_1 - a$.

1.4   Properties of Convergent Series

a. **Cauchy's Criterion**: Series $\sum_{k=1}^\infty a_k$ converges if and only if

$$\forall \varepsilon > 0, \ \exists N(\varepsilon), \ \forall n > N(\varepsilon), \ \forall p \in \mathbb{N}$$

$$|a_{n+1} + a_{n+2} + \ldots + a_{n+p}| = \left| \sum_{k=n+1}^{n+p} a_k \right| < \varepsilon$$

b. **Necessary Condition of Convergence**

If series $\sum_{k=1}^\infty a_k$ converges, then $\lim_{n\to\infty} a_n = 0$.

This is an **insufficient condition**: For instance, for harmonic series $\sum_{n=1}^\infty \frac{1}{n}$, there holds $\lim_{n\to\infty} \frac{1}{n} = 0$, yet the series is divergent.

c. If series $\sum_{n=1}^\infty a_n$ is convergent, then the sequence of remainders $\{R_n\}$ converges to zero. That is, for every

$\varepsilon > 0$ there exists $N_0(\varepsilon)$ such that for all $n > N_0(\varepsilon)$, there holds $|R_n| < \varepsilon$.

1.5   Operation on Series

a. Removing a finite number of terms from a series, or adding a finite number of terms to it, does not affect the convergence or divergence of the series.

Attention: It does change the sum of the series.

b. If $\displaystyle\sum_{k=1}^{\infty} a_k$ converges to S, then, for every constant c the series $\displaystyle\sum_{k=1}^{\infty} ca_k$ converges to $c \cdot S$.

If series $\displaystyle\sum_{k=1}^{\infty} a_k$ is divergent, then, for all $c \neq 0$, the series $\displaystyle\sum_{k=1}^{\infty} ca_k$, is divergent.

c. If series $B = \displaystyle\sum_{k=1}^{\infty} b_k$, $A = \displaystyle\sum_{k=1}^{\infty} a_k$ are convergent, then, the series $\displaystyle\sum_{k=1}^{\infty} (a_k \pm b_k)$ is convergent, and

$$\sum_{k=1}^{\infty} (a_k \pm b_k) = \sum_{k=1}^{\infty} a_k \pm \sum_{k=1}^{\infty} b_k = A \pm B.$$

Chapter 2
Positive Series

A positive series is a series of all-positive terms, except, possibly, a finite number of terms.

2.1   Comparison Tests

Given two positive series (A) $\sum_{n=1}^{\infty} a_n$; (B) $\sum_{n=1}^{\infty} b_n$.

a. **First Comparison Test**: if, for all n starting from a certain number $a_n \le b_n$, then:

1. From the convergence of series (B), there follows the convergence of series (A).

2. From the divergence of series (A), there follows the divergence of series (B).

b. **Second Comparison Test**: Let the limit $\lim_{n \to \infty} \dfrac{a_n}{b_n} = k$ exists:

1. If $0 < k < \infty$, then series (A) and (B) are either both convergent or both divergent.

2 If $k = 0$, then from the convergence of series (B), there follows the convergence of series (A).

3. If $k = \infty$, then from the convergence of series (A), there follows the convergence of series (B).

c. **Third Comparison Test**: if, for all n starting from a certain number

$$\frac{a_{n+1}}{a_n} \le \frac{b_{n+1}}{b_n}, \quad (a_n \neq 0, b_n \neq 0)$$

then:

1. From the convergence of series (B), there follows the convergence of series (A).

2. From the convergence of series (A), there follows the convergence of series (B).

2.2   D'Alembert's Ratio Test

a. If, starting from a certain n, for the terms of series, $\sum_{n=1}^{\infty} a_n$, $a_n > 0$, there holds:

1. $\frac{a_{n+1}}{a_n} \le q < 1$, then the series is convergent.

2. $\frac{a_{n+1}}{a_n} \ge 1$, then the series is divergent.

b. **The Limit Form of Ratio Test**: if the limit $\lim_{n \to \infty} \frac{a_{n+1}}{a_n} = L$ exists, then, when:

1. $L < 1$, the series is convergent.

2. $L > 1$, the series is divergent.

3. $L = 1$, the test is inconclusive regarding the convergence or divergence of the series.

2.3   Cauchy's Root Test

a. If, starting from a certain n, for the terms of series, $\sum\limits_{n=1}^{\infty} a_n$,

   $a_n \geq 0$, there holds:

   1. $\sqrt[n]{a_n} \leq q < 1$, then the series is convergent,

   2. $\sqrt[n]{a_n} \geq 1$, then the series is divergent.

b. **The Limit Form of Root Test**: if the limit $\lim\limits_{n\to\infty} \sqrt[n]{a_n} = L$

   exists, then, when:

   1. $L < 1$, the series is convergent.

   2. $L > 1$, the series is divergent.

   3. $L = 1$, the test is inconclusive regarding the convergence or divergence of the series.

2.4   Integral Test

a. If $f(x)$ is a positive function, non-increasing in domain $x \geq 1$, and a_n equals the value of function at $x = n$, then

   the series $\sum\limits_{n=1}^{\infty} a_n = \sum\limits_{n=1}^{\infty} f(n)$ and integral $\int\limits_{1}^{\infty} f(x)dx$ are either

   both convergent or both divergent (see IV, 3.1).

b. If a series is convergent, then, for its remainder R_n , there holds

$$\int\limits_{n+1}^{\infty} f(x)dx \leq R_n \leq a_{n+1} + \int\limits_{n+1}^{\infty} f(x)dx$$

Chapter 3
General Series

Series containing an infinite number of positive terms and an infinite number of negative terms are called **general series**.

3.1   Absolute and Conditional Convergence

a. Series $\sum\limits_{n=1}^{\infty} a_n$ is **absolutely convergent** if series $\sum\limits_{n=1}^{\infty} |a_n|$ is convergent.

b. An absolutely convergent series is convergent.

c. A convergent, but not absolutely convergent series, is **conditionally convergent**.

d. **Examples**:

1. The series $\sum\limits_{n=1}^{\infty} \dfrac{(-1)^n}{n^2}$ is absolutely convergent, since series $\sum\limits_{n=1}^{\infty} \left| \dfrac{(-1)^n}{n^2} \right| = \sum\limits_{n=1}^{\infty} \dfrac{1}{n^2}$ is convergent.

2. Leibniz series $\sum\limits_{n=1}^{\infty} \dfrac{(-1)^{n+1}}{n}$ is conditionally convergent since the series itself is convergent and series $\sum\limits_{n=1}^{\infty} \left| \dfrac{(-1)^{n+1}}{n} \right| = \sum\limits_{n=1}^{\infty} \dfrac{1}{n}$ is divergent.

3.2   Operations on the Terms of a Series

a. If series $\sum_{n=1}^{\infty} a_n$ converges, then, all series resulting from grouping its terms in sets, without changing the order of terms from the original series, converges to the same sum.

The inverse is incorrect. That is, if a series with groups is convergent, the series without groups is not necessarily convergent.

Example: The series $(1-1)+(1-1)+...$ is convergent, yet the ungrouped series $1-1+1-1+...$ is divergent.

b. **Cauchy's Theorem**

If series $\sum_{n=1}^{\infty} a_n$ is absolutely convergent, then, all series resulting from any changer of the order of its terms absolutely converges to the same sum.

c. **Riemann Theorem**

If series $\sum_{n=1}^{\infty} a_n$ is conditionally convergent, then its terms can be arranged in such a way that the rearranged series will converge to a predetermined sum L, or even diverge.

Example: Let's rearrange the terms of a Leibniz series

$$\ln 2 = \sum_{n=1}^{\infty} \frac{(-1)^{n+1}}{n} = 1 - \frac{1}{2} + \frac{1}{3} - \frac{1}{4} + ... + \frac{1}{2k-1} - \frac{1}{2k} + ...$$

the following way:

$$1 - \frac{1}{2} - \frac{1}{4} + \frac{1}{3} - \frac{1}{6} - \frac{1}{8} + ... + \frac{1}{2k-1} - \frac{1}{2k-2} - \frac{1}{4k} + ...$$

We denote by S_n and S'_n the partial sums of the Leibniz series and the rearrange series, respectively:

$$S'_{3n} = \left[\left(1 - \frac{1}{2}\right) - \frac{1}{4}\right] + \left[\left(\frac{1}{3} - \frac{1}{6}\right) - \frac{1}{8}\right] + \ldots + \left[\left(\frac{1}{2n-1} - \frac{1}{4n-2}\right) - \frac{1}{4n}\right] =$$

$$= \frac{1}{2} - \frac{1}{4} + \frac{1}{6} - \frac{1}{8} + \ldots + \frac{1}{4n-2} - \frac{1}{4n} =$$

$$= \frac{1}{2}\left(1 - \frac{1}{2} + \frac{1}{3} - \frac{1}{4} + \ldots - \frac{1}{2n}\right) = \frac{1}{2}S_{2n} \xrightarrow[n \to \infty]{} \frac{1}{2}\ln 2$$

Therefore, the rearrange series converges to a different sum, $\frac{1}{2}\ln 2$.

3.3   General Series Convergence Tests

a. Leibniz Test

If sequence $\{a_n\}$ is monotonic, positive, and $\lim\limits_{n \to \infty} a_n = 0$, then, the sign-changing series $\sum\limits_{n=1}^{\infty}(-1)^{n+1}a_n$ is:

1. Convergent.

2. For its sum S, there holds: $0 < S < a_1$.

3. Its remainder, $|R_n| < a_{n+1}$.

b. Dirichlet Test

If any term of the series $\sum\limits_{n=1}^{\infty}u_n$ can be presented as the product of two terms, $u_n = a_n \cdot b_n$, such that:

1. Sequence $\{a_n\}$ is monotonic and converges to zero.

2. All partial sums of series $\sum_{n=1}^{\infty} b_n$ are common bounded, that is, there exists number M such that for all n, there holds $|S_n| \le M$, then series $\sum_{n=1}^{\infty} u_n$ is convergent.

c. **Abel's Test**

If any term of the series $\sum_{n=1}^{\infty} u_n$ can be presented as the product of two terms, $u_n = a_n \cdot b_n$, such that:

1. Sequence $\{a_n\}$ is monotonic and bounded.

2. Series $\sum_{n=1}^{\infty} b_n$ is even conditionally convergent, then series $\sum_{n=1}^{\infty} u_n$ is convergent.

3.4   Product of Series

a. A series consisting of all products $a_i b_j$ when $i, j = 1, 2, \ldots$ is called the product of series $A = \sum_{n=1}^{\infty} a_n$, $B = \sum_{n=1}^{\infty} b_n$.

b. If both series are absolutely convergent, then their product absolutely converges to $A \cdot B$ and:

$$\left(\sum_{n=1}^{\infty} a_n \right) \left(\sum_{n=1}^{\infty} b_n \right) = \sum_{n=1}^{\infty} (a_1 b_n + a_2 b_{n-1} + \ldots + a_n b_1)$$

Chapter 4
Series of Functions

4.1 Sequences of Functions

a. $f_1(x), f_2(x), \ldots, f_n(x), \ldots, x \in D$ is a sequence of functions, and D is their domain.

b. The sequence of functions $\{f_n(x)\}$ converges at x_0 if the sequence $\{f_n(x_0)\}$ converges to $f(x_0)$. Written:

$$\lim_{n \to \infty} f_n(x_0) = f(x_0)$$

c. Set I of all points $x_0 \in D$ where the sequence of functions converges, is called **domain of convergence** of the sequence $I \subset D$.

d. For all x_0 of set I, there is a target value $f(x_0)$. All limit values define the function $x \in I$, $f(x)$ called **limit function**.

e. The sequence of functions $\{f_n(x)\}$ **uniformly converges** to function $f(x)$ in I, if for every $\varepsilon > 0$, there exists $N(\varepsilon)$ (dependent of ε only), such that for all $n > N(\varepsilon)$ and for all $x \in I$, there holds $|f_n(x) - f(x)| < \varepsilon$.

f. **Criteria of Uniform Convergence**

Cauchy's Criterion: sequence $\{f_n(x)\}$ uniformly converges in domain D if and only if there exists $N(\varepsilon)$ (dependent of ε only), such that for all $n > N(\varepsilon)$ and for all $x \in I$, and for all integer p, there holds $|f_{n+p}(x) - f_n(x)| < \varepsilon$.

g. Sequence of functions $\{f_n(x)\}$ uniformly converges to $f(x)$ in domain D if and only if $\lim\limits_{n\to\infty} \sup\limits_{x\in I} |f_n(x) - f(x)| = 0$.

h. If a sequence of continuous functions $\{f_n(x)\}$ in D uniformly converges in D to $f(x)$, then $f(x)$ is continuous in D.

4.2 Series of Function

a. Series $\sum\limits_{n=1}^{\infty} f_n(x) = f_1(x) + f_2(x) + ...$, when $\{f_n(x)\}$ is a sequence of factions defined in common domain D is a **series of functions**.

b. At a constant $x = x_0$, the series is a series of numbers $\sum\limits_{n=1}^{\infty} f_n(x_0)$.

c. The set of all points on D where the series converges is called the **domain of convergence** of the series.

d. The **sum of the series**, $S(x)$ is a function defined in the series domain of convergence, and $S(x) = \sum\limits_{n=1}^{\infty} f_n(x)$.

4.3 Uniform Convergence of Series

a. The series $\sum\limits_{n=1}^{\infty} f_n(x)$ is uniformly convergent in D if the sequence of its partial sums is uniformly convergent in D.

b. **Cauchy's Criterion**: a series is uniformly convergent in D if and only if there exists $N(\varepsilon)$ (dependent of ε only), such that for every $\varepsilon > 0$ for all $n > N(\varepsilon)$, for all integer p and for all $x \in E_0$, there holds $\left| \sum\limits_{k=n+1}^{n+p} f_k(x) \right| < \varepsilon$.

c. **Weierstrass Test**: if for series of functions $\sum\limits_{n=1}^{\infty} f_n(x)$ defined in D there exists a positive convergent series $\sum\limits_{n=1}^{\infty} a_n$ such that, starting from some n, $|f_n(x)| \leq a_n$ for all $x \in D$, then the series is uniformly convergent in D.

d. Given series $\sum\limits_{n=1}^{\infty} a_n(x) \cdot b_n(x)$.

    1. **Dirichlet Test**: if all partial sums $B_n(x)$ of the series $\sum\limits_{n=1}^{\infty} b_n(x)$ have common bound, that is, there exists M such that $|B_n(x)| < M$ for all $x \in D$ and all n, and the sequence $\{a_n(x)\}$ is monotonic and uniformly convergent to zero at D, then the given series is uniformly convergent in D.

    2. **Abel's Test**: if the sequence $\{a_n(x)\}$ is monotonic and bounded in D, and the series $\sum\limits_{n=1}^{\infty} b_n(x)$ is uniformly convergent in D, then the given series, also, is uniformly convergent in D.

4.4   Continuity, Derivability and Integrability of Sums of Series

a. If functions $f_n(x)$, $(n = 1, 2, \ldots)$ are continuous in domain D and the series $S(x) = \sum\limits_{n=1}^{\infty} f_n(x)$ is uniformly convergent in D, then function $S(x)$ is continuous in D.

b. If the sum of a series of continuous functions converges to a discontinuous function in the same domain, then the convergence is not uniform.

c. If $\{f_n(x)\}$ is a series of functions continuous in $[a,b]$, and the series of functions $S(x) = \sum\limits_{n=1}^{\infty} f_n(x)$ uniformly converges to $S(x)$ on $[a,b]$, then:

$$\int_a^b S(x)\,dx = \int_a^b \left(\sum_{n=1}^{\infty} f_n(x) \right) dx = \sum_{n=1}^{\infty} \int_a^b f_n(x)\,dx$$

d. If functions $f_n(x)$, $(n = 1,2,...)$ are derivable and have continuous derivatives on interval $[a,b]$, the series $S(x) = \sum\limits_{n=1}^{\infty} f_n(x)$ is convergent on $[a,b]$ and series of derivatives $\sum\limits_{n=1}^{\infty} f'_n(x)$ is uniformly convergent on $[a,b]$, then:

$$S'(x) = \left(\sum_{k=1}^{\infty} f_k(x) \right)' = \sum_{k=1}^{\infty} f'_k(x)$$

In other words, the derivative of a sum equals the sum of derivatives.

4.5   Power Series and Radius of Convergence

A functional series in the form of

$$\sum_{n=0}^{\infty} a_n x^n = a_0 + a_1 x + a_2 x^2 + ..., \text{ or}$$

$$\sum_{n=0}^{\infty} a_n (x - x_0)^n = a_0 + a_1(x - x_0) + a_2(x - x_0)^2 + ... \text{ is called a}$$

power series.

Substituting $x - x_0 = t$ in the latter series, will result in the former series.

a. Radius of Convergence Existence Theorem

For all power series $\sum\limits_{n=0}^{\infty} a_n x^n$ there exists non-negative R, $(0 \leq R \leq \infty)$, such that for all x which holds $|x| < R$, the series is convergent, and for all x which holds $|x| > R$, the series is divergent, and if $R = 0$, the series converges at $x = 0$ only. If $R = \infty$, the series converges for all $-\infty < x < \infty$. R is called the **radius of convergence** of the series.

b. Formulas for Calculating the Radius of Convergence

1. $R = \lim\limits_{n \to \infty} \left| \dfrac{a_n}{a_{n+1}} \right|$

2. Cauchy-Hadamard formula: $R = \dfrac{1}{\lim\limits_{n \to \infty} \sqrt[n]{|a_n|}}$

4.6 Uniform Convergence, Derivation and Integration of Power Series

a. If power series $\sum\limits_{n=0}^{\infty} a_n x^n$ with a convergence radius $R > 0$:

1. Diverges at endpoint $x = R$ $(x = -R)$, then the convergence on interval $(-R, R)$ is not uniform.

2. Converges at endpoint $x = R$, $(x = -R)$, then the convergence is uniform on intervals $[0, R]$, $([-R, 0])$.

b. The series $\sum\limits_{n=0}^{\infty} a_n x^n$ is uniformly convergent on every segment $[\alpha, \beta]$, which is entirely within $(-R, R)$.

c. Sum $f(x)$ of power series with radius of convergence $R > 0$ is a function continuous on all $x \in (-R, R)$. If, in addition, it converges on $x = R$, $(x = -R)$, then $f(x)$ is continuous from the left (from the right) at endpoint $x = R$, $(x = -R)$.

d. Let $R > 0$ be the radius of convergence of a power series $f(x) = \sum_{k=0}^{\infty} a_k x^k$. Then, for all $|x| < R$, there holds:

1. $\int_0^x f(t)\, dt = \int_0^x \left(\sum_{k=0}^{\infty} a_k t^k \right) dt = \sum_{k=0}^{\infty} \frac{a_k}{k+1} x^{k+1}$.

2. Both series have the same radius of convergence R.

3. If a power series converges at $x = R$, $(x = -R)$, then, their integrals series also converges at $x = R$, $(x = -R)$.

e. If Let $R > 0$ be the radius of convergence of a power series. Then, for all $|x| < R$, there holds:

1. $f'(x) = \left(\sum_{k=0}^{\infty} a_k x^k \right)' = \sum_{k=1}^{\infty} k a_k x^{k-1}$

2. The power series and derivatives series have the same power of convergence R.

3. If the series of derivatives converges at $x = R$, $(x = -R)$, then, the original series converges at the same endpoint.

4.7   Power Series Expansion of Functions

Function $f(x)$ defined on $(-R, R)$ is expanded to a **power Taylor series**, if there exists a power series converging to $f(x)$ on $(-R, R)$.

a. **A Necessary Condition of Expansion to Power Series**:
if $f(x)$ is the sum of series $\sum\limits_{n=0}^{\infty} a_n x^n$ on $(-R,R)$, then
$f(x)$ is infinitely derivable and all its derivatives are
functions contiguous on $(-R,R)$.

b. **The Expansion of $f(x)$ to Power (Binomial) Series in Powers of** $x - x_0$:

$$f(x) = f(x_0) + f'(x_0)(x - x_0) + \frac{f''(x_0)}{2!}(x - x_0)^2 + \frac{f^{(3)}(x_0)}{3!}(x - x_0)^3 + \ldots$$

converging at $(x_0 - R, \ x_0 + R)$.

c. **Examples**:

1. $e^x = 1 + x + \dfrac{x^2}{2!} + \ldots + \dfrac{x^n}{n!} + \ldots = \sum\limits_{n=0}^{\infty} \dfrac{x^n}{n!}$, $|x| < \infty$

2. $\sin x = x - \dfrac{x^3}{3!} + \ldots + \dfrac{(-1)^n}{(2n+1)!} x^{2n+1} + \ldots$

$\qquad = \sum\limits_{n=0}^{\infty} \dfrac{(-1)^n x^{2n+1}}{(2n+1)!}$, $|x| < \infty$

3. $\cos x = 1 - \dfrac{x^2}{2!} + \dfrac{x^4}{4!} - \ldots + \dfrac{(-1)^n x^{2n}}{(2n)!} + \ldots$

$\qquad = \sum\limits_{n=0}^{\infty} \dfrac{(-1)^n x^{2n}}{(2n)!}$, $|x| < \infty$

4. $\ln(1+x) = x - \dfrac{x^2}{2} + \dfrac{x^3}{3} - \ldots = \sum\limits_{n=1}^{\infty} \dfrac{(-1)^{n+1}}{n} x^n$, $-1 < x \leq 1$

5. $\dfrac{1}{1+x} = 1 - x + x^2 - \ldots = \sum\limits_{n=0}^{\infty} (-1)^n x^n$, $|x| < 1$

6. $\dfrac{1}{1-x} = 1 + x + x^2 + \ldots = \sum_{n=0}^{\infty} x^n$, $|x| < 1$

7. $(1+x)^{\alpha} = 1 + \sum_{k=1}^{\infty} \dfrac{\alpha(\alpha-1)\ldots(\alpha-k+1)}{k!} x^k$, $|x| < 1$

VII. Differential Calculus of Multivariable Functions

Chapter 1
Introduction

1.1   Domain in $\mathbb{R}^n$

a. $M(x_1, x_2, \ldots, x_n), x_i \in \mathbb{R}, 1 \le i \le n$ is a point on $\mathbb{R}^n$. x_i are its coordinates.

b.   The distance between $M_1(x_1^{(1)}, x_2^{(1)}, \ldots, x_n^{(1)})$ and

$M_2(x_1^{(2)}, x_2^{(2)}, \ldots, x_n^{(2)})$ is $d(M_1, M_2) = \sqrt{\sum_{i=1}^{n}[x_i^{(1)} - x_i^{(2)}]^2}$.

c.   An n -**dimensional sphere** of radius R, with its center at $M_0(x_1^{(0)}, x_2^{(0)}, \ldots, x_n^{(0)})$ is the locus of all points $(x_1, x_2, \ldots, x_n) \in \mathbb{R}^n$ the distance of which from M is smaller than or equal to R, and which hold: $(x_1 - x_1^{(0)})^2 + (x_2 - x_2^{(0)})^2 + \ldots + (x_n - x_n^{(0)})^2 \le R^2$.

d.   A sphere is an **open sphere** if this is a strict inequality.

e. $\varepsilon > 0$ **neighborhood** of M_0 is an open sphere with radius ε centered at M_0.

f. M_0 is an **interior point** of set $\{M\}$ of $\mathbb{R}^n$ if there exists ε -neighborhood of M_0 which is entirely at $\{M\}$.

g. $\{M\}$ is an **open set** in $\mathbb{R}^n$ if all is points are interior points.

h. Set $\{M\}$ is **bounded** if there exists a sphere of a finite radius containing it.

i. **Continuous line** at $\mathbb{R}^n$ is the locus of all points $(x_1,...,x_n)$ of $x_i = \varphi_i(t)$, $\alpha < t < \beta$, $(i = 1, 2, ..., n)$ coordinates when $\varphi_i(t)$ are continuous functions of parameter t .

j. D is a **connected set** if every two interior points of it can be connected by a continuous line which is entirely at D .

k. An open and connected set is called a **domain**.

1.2   Sequences of Points

a. **Limit Point of Sequence**

$\{M_k\}_{k=1}^{\infty} = \left\{(x_1^{(k)}, x_2^{(k)}, ..., x_n^{(k)})\right\}_{k=1}^{\infty}$ is point $A(a_1, a_2, ..., a_n)$, if for every $\varepsilon > 0$ there exists $n_0(\varepsilon)$ such that for all $k > n_0(\varepsilon)$ there holds $d(M_k, A) < \varepsilon$. It is denoted $\lim_{k \to \infty} M_k = A$.

b. Set of points $\{M_k\}_{k=1}^{\infty} = (x_1^{(k)}, x_2^{(k)}, ..., x_n^{(k)})\}$ converges to A if and only if n sequences of coordinates $\{x_1^{(k)}\}, \{x_2^{(k)}\}, ..., \{x_n^{(k)}\}$ converge to coordinates $a_1, a_2, ..., a_n$, respectively.

c. **Bolzano-Weierstrass Theorem**: every bounded and infinite sequence of points at $\mathbb{R}^n$ has a subset converging to the limit.

Chapter 2
Multivariable Function

2.1   Definition

a. Let D be a set at $\mathbb{R}^n$. A function $f:\{D\} \to \mathbb{R}^1$ is a given rule fitting each point M of D fits a one and only number $u \in \mathbb{R}^1$. It is denoted $u = f(M)$ or $u = f(x_1, \ldots, x_k)$.

   Set $\{D\}$ is the **domain** of the function.

b. The set of all values u is called the **range** of the function.

c. The geometric description of a function of two-variables $z = f(x,y)$ is a surface at $\mathbb{R}^3$ (see III.8 for examples of such surfaces).

2.2   Level Curve and Level Surface

a. **Level curve**: Curve L at plane x,y is called a **level curve** or a **level set** of function $z = f(x,y)$, fitting constant c, if the value of the function on L equals c. That is, L is the locus of points (x,y) which hold $f(x,y) = c$.

   The intersection between plane $z = c$ and surface $f(x,y)$ is curve L.

   Figure 1 shows level curves of function $z = \sqrt{x^2 + y^2}$, which are circles $x^2 + y^2 = c^2, c > 0$.

b. Surface S is called a **level surface** fitting constant c, of function $u = f(x,y,z)$, if it is defined the equation $f(x,y,z) = c$.

Figure 2 shows level surfaces of function $f(x,y,z) = x^2 + y^2 - z$, when $c = 0$, $c = 10$, $c = 25$.

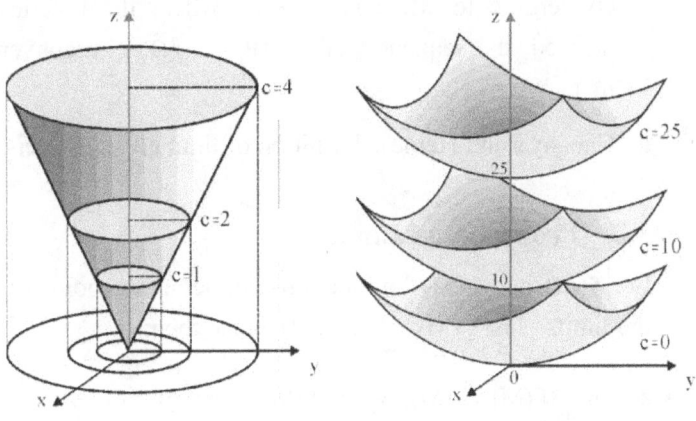

Figure 1   Figure 2

Chapter 3
Limits and Continuity of Functions

3.1   Definition of Limit

a. **Cauchy's Definition**: L is the limit of function $f(M)$ at the point $M \to M_0$, if for every $\varepsilon > 0$ there exists $\delta > 0$, such that for all M holding $0 < d(M, M_0) < \delta$, there holds $|f(M) - L| < \varepsilon$. It is written $\lim\limits_{M \to M_0} f(x) = L$.

b. Limit L is not dependent of the path through which point M tends to M_0.

c. **Heine's Definition**: L is the **limit of function** $u = f(M)$ when $M \to M_0$, if for all sequences of points $\{P_n\}$ converging to M_0 and $P_n \neq M_0$ where the function is defined, the sequence $f(P_1), f(P_2), \ldots, f(P_n), \ldots$ converges to L.

d. Cauchy's and Heine's definition of limit are equivalent.

3.2   Properties of Limit

Let $f(M)$ and $g(M)$ be functions defined at the point M_0. If the limits $\lim\limits_{M \to M_0} g(M)$, $\lim\limits_{M \to M_0} f(M)$ exist, then:

a. $\lim\limits_{M \to M_0} [f(M) \pm g(M)] = \lim\limits_{M \to M_0} f(M) \pm \lim\limits_{M \to M_0} g(M)$

b. $\lim\limits_{M \to M_0} [f(M) \cdot g(M)] = \lim\limits_{M \to M_0} f(M) \cdot \lim\limits_{M \to M_0} g(M)$

c. If, in addition, $g(M) \neq 0$ and $\lim\limits_{M \to M_0} g(M) \neq 0$, then:

$$\lim_{M \to M_0} \frac{f(M)}{g(M)} = \frac{\lim\limits_{M \to M_0} f(M)}{\lim\limits_{M \to M_0} g(M)}.$$

3.3   Continuity at a Point

Function $u = f(M)$ is **continuous at point** M_0 if for every $\varepsilon > 0$ there exists $\delta > 0$ such that for all M holding $d(M, M_0) < \delta$ there holds $|f(M) - f(M_0)| < \varepsilon$.

In other words, function $u = f(M)$ is continuous at $M_0 \in D$ if $\lim_{M \to M_0} f(M) = f(M_0)$.

3.4   Properties of Continuous Functions

a. If function $u = f(M)$ is continuous on $M_0 \in D \subset \mathbb{R}^k$ and $f(M_0) > 0$, $(f(M_0) < 0)$, then there exists a neighborhood of M_0 such that at all point M of that neighborhood, $f(M) > 0$, $(f(M) < 0)$.

b. If functions $f(M)$ and $g(M)$ are continuous on M_0, then:

1. Functions $f(M) \pm g(M)$, $f(M) \cdot g(M)$ are continuous on M_0.

2. If, in addition, $g(M_0) \neq 0$, then, function $\dfrac{f(M)}{g(M)}$, is continuous on M_0.

c. **Continuity of a Composite Function**

Theorem: Let $f(M) = f(x_1, x_2, ..., x_k)$ be a function defined by $D \subset \mathbb{R}^k$, and functions

$$x_1 = \varphi_1(t_1, ..., t_m), x_2 = \varphi_2(t_1, ..., t_m), ..., x_k = \varphi_k(t_1, ..., t_m) \quad (*)$$

defined by $\Delta \subset \mathbb{R}^m$ and let $M(x_1, x_2, ..., x_k)$ be a point on D and $P(t_1, t_2, ..., t_m)$ a point on Δ, the coordinates of which are connected by (*).

If functions $\varphi_i(P)$, $(i = 1, 2, ..., k)$ are continuous on $P_0(t_1^0, ..., t_m^0) \in \Delta$ and function $f(x_1, x_2, ..., x_k)$ is continuous on $M_0(x_1^0, x_2^0, ..., x_k^0) \in D$, such that $x_i^0 = \varphi_i(t_1^0, t_2^0, ..., t_m^0)$, $i = 1, 2, ..., k$, then, the composite function $f(\varphi_1(P), \varphi_2(P), ..., \varphi_k(P))$ is continuous on P_0.

In other words, a **composition of continuous functions is a continuous function**.

d. Function $u = f(M)$ is **continuous on domain** D if it is continuous on all points of D.

e. **Intermediate Value Theorem**: If function $u = f(M)$ is continuous in connected domain $D \subset \mathbb{R}^k$, and if points A,B are on D, then, for all real number m between $f(A)$ and $f(B)$ there exists point C such that $f(C) = m$.

f. **Weierstrass Theorem**: If function $u = f(M)$ is continuous on closed and bounded domain D, then it is bounded on that domain, reaching its maximum and minimum value above D. That is, there exist points A, B on D, such that:

$$\min_{M \in D} f(M) = f(B) \ , \ \max_{M \in D} f(M) = f(A)$$

3.5   Uniform Continuity

a. Function $u = f(M)$ is continuous on domain D if, for every $\varepsilon > 0$ there is $\delta > 0$, dependent on ε, such that for all two points M",M' which hold $d(M',M") < \delta$, there holds $|f(M") - f(M')| < \varepsilon$.

b. **Cantor Theorem**: If function $u = f(M)$ is continuous on closed and bounded domain D, then it is uniformly continuous on that domain.

Chapter 4
Partial Derivatives

4.1   Definition of Partial Derivatives

Let function $u = f(x_1, x_2, ..., x_k)$ be defined on domain D and point $M_0 = (x_1^0, x_2^0, ..., x_k^0)$ on D.

Selecting variables $x_2, x_3, ..., x_k$, we construct a function $u = f(x_1, x_2^0, ..., x_k^0)$ of one variable, x_1.

Adding to x_1 a Δx_1 at x_1^0, we get addition $\Delta_1 u$:

$$\Delta_1 u = f(x_1^0 + \Delta x_1, x_2^0, ..., x_k^0) - f(x_1^0, x_2^0, ..., x_k^0)$$

If the limit $\lim\limits_{\Delta x_1 \to 0} \dfrac{\Delta_1 u}{\Delta x_1}$ exists, it is called the **partial derivative** of function $u = f(M)$ with respect to x_1 on M_0. It is denoted as either, $f_{x_1}(M_0)$, $f'_{x_1}(M_0)$, $u_{x_1}(M_0)$, $\dfrac{\partial f}{\partial x_1}(M_0)$ or $\dfrac{\partial u}{\partial x_1}(M_0)$.

Similarly, we defined partial derivatives with respect to $x_2, x_3, ..., x_k$

4.2   Geometric Description of Partial Derivatives

Let S be a surface defined by the function $z = f(x, y)$, and let point (x_0, y_0, z_0) be on surface S when $z_0 = f(x_0, y_0)$.

Fixing $y = y_0$, we get a one-variable function $z = f(x, y_0)$ defining the line of intersection between plane $y = y_0$ and surface S. Therefore, $\dfrac{\partial z}{\partial x} = f_x(x_0, y_0)$ defines the slope of

tangent line m at point (x_0, y_0, z_0) to curve $z = f(x, y_0)$ and $\tan\alpha = f_x(x_0, y_0)$ (Figure 1).

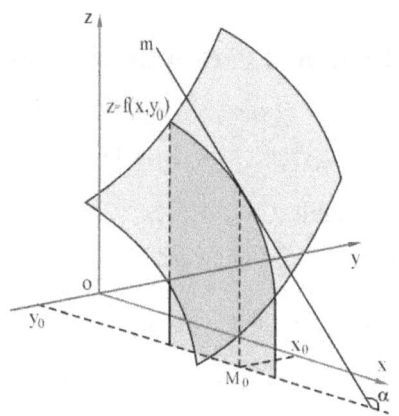

Figure 1

That is to say, $f_x(x_0, y_0)$ describes the rate of change of function $z = f(x, y)$ towards the x-axis. Similarly, $f_y(x_0, y_0)$ describes the rate of change of function $z = f(x, y)$ towards the y-axis, and equals $\tan\beta$ which is the slope of tangent line n (Figure 2).

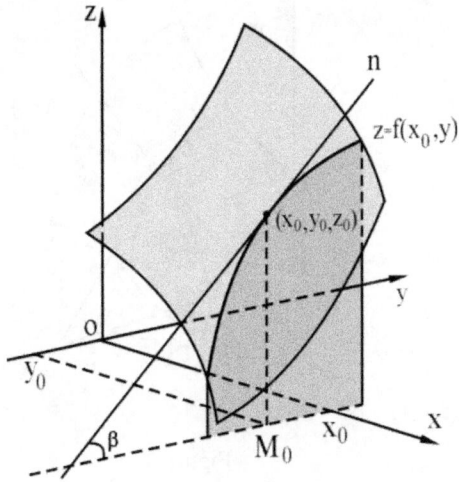

Figure 2

The equation of tangent line m to curve $z = f(x, y_0)$ at $M_0(x_0, y_0, z_0)$ is

$$z - z_0 = f_x(x_0, y_0) \cdot (x - x_0)$$

The equation of tangent line n to curve $z = f(x_0, y)$ at point (x_0, y_0, z_0) is

$$z - z_0 = f_y(x_0, y_0) \cdot (y - y_0)$$

Through these tangent lines, passes a plane expressed by the equation

$$z - z_0 = f_x(x_0, y_0)(x - x_0) + f_y(x_0, y_0)(y - y_0)$$

This is a **tangent plane** to surface $z = f(x, y)$ at point M_0. Vector $\left(f_x(x_0, y_0), f_y(x_0, y_0), -1 \right)$ is the **normal** to the surface at M_0 (Figure 3).

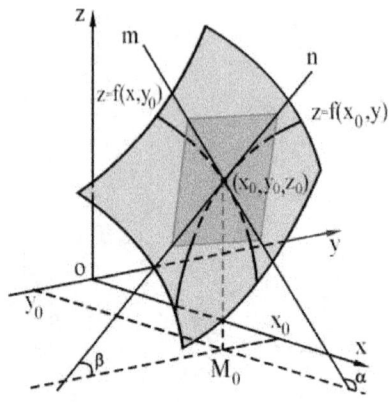

Figure 3

4.3   Higher-Order Partial Derivatives

a. Partial   derivative $\dfrac{\partial u}{\partial x_i}$ by x_i of   function $u = f(x_1, x_2, \ldots, x_k)$, is also a function of k variables. If it is partially differentiable on x_j, the result is second-order partial derivatives, denoted by:

$$f_{ij}^{(2)},\ f_{x_i x_j},\ f_{x_i, x_j}^{(2)}\ \text{or}\ \frac{\partial^2 u}{\partial x_j \partial x_i},\ (i, j = 1, 2, \ldots, k)$$

If $i \neq j$, derivative $f_{ij}^{(2)}$ is called a second-order mixed **partial derivative**.

b. If function $f(x, y)$ is defined on domain D and has partial derivatives f_{yx}, f_{xy}, f_y, f_x continuous in the neighborhood of $M_0(x_0, y_0) \in D$, then, at that point, the mixed partial derivatives are equal: $f_{xy}(x_0, y_0) = f_{yx}(x_0, y_0)$.

c. Function $f(x_1, x_2, ..., x_k)$ is said to be of to **class** C on domain D if it is continuous on D. Function $f(M)$ belongs to **class** C^n if it is continuous and has continuous partial derivatives up to and including n-th order, on D.

d. If function $u = f(x_1, x_2, ..., x_k)$ is defined on k-dimensional domain D and belongs to class C^n, then the value of n-th order mixed derivatives is independent of the order of derivation.

Chapter 5
Differentiability. Taylor's Formula

5.1   Differentiability of Functions of Two-Variable

a. Function $z = f(x, y)$ is differentiable at point (x_0, y_0), if its general addition $\Delta z = f(x_0 + \Delta x, y_0 + \Delta y) - f(x_0, y_0)$ can be represented the following way:

$$\Delta z = f_x(x_0, y_0) \Delta x + f_y(x_0, y_0) \Delta y + \varepsilon \sqrt{\Delta x^2 + \Delta y^2}$$

when $\lim_{\Delta x \to 0, \Delta y \to 0} \varepsilon = 0$.

b. If function $z = f(x, y)$ is differentiable at point M_0, then it is continuous at that point.

The inverse proposition is incorrect.

c. If function $z = f(x, y)$ is defined on the neighborhood of $M_0(x_0, y_0)$, and has partial derivatives $f_x(x, y)$, $f_y(x, y)$ continuous in that neighborhood, then it is differentiable at M_0.

5.2   Differentiability of Function $u = f(x_1, x_2, ..., x_k)$

a. Let Δx_i be an addition to x_i, $(i = 1, 2, ..., k)$, and let

$$\Delta u = f\left(x^0_1 + \Delta x_1, \ x^0_2 + \Delta x_2, ...x^0_k + \Delta x_k\right) - f\left(x^0_1, \ x^0_2, ...x^0_k\right)$$

be an addition to function at $M_0(x^0_1, x^0_2, ..., x^0_k)$.

If addition $\Delta u = f(M) - f(M_0)$ can be represented in the following way:

$$\Delta u = \sum_{i=1}^{k} f_{x_i}(M_0) \Delta x_i + \varepsilon \sqrt{\sum_{i=1}^{k} (\Delta x_i)^2}$$

when $\lim_{\substack{\Delta x_1 \to 0 \\ \vdots \\ \Delta x_k \to 0}} \varepsilon = 0$, $(i = 1, 2, ..., k)$, then function u is said to be **differentiable** on M_0.

b. If function $f(x_1, x_2, ..., x_k)$ is continuous on M_0, and has partial derivatives continuous in the neighborhood of M_0, then it is differentiable at M_0.

5.3   Differential

a. The linear part of Δu in relation to Δx_i is the differential of u and is denoted as du :

$$du = \frac{\partial u}{\partial x_1}(M_0) dx_1 + \frac{\partial u}{\partial x_2}(M_0) dx_2 + ... + \frac{\partial u}{\partial x_k}(M_0) dx_k$$

when $dx_i = \Delta x_i$.

b. The differential of two-variable function $u = f(x, y)$ is

$$du = f_x dx + f_y dy$$

c. Second-order differential:

$$d(du) = d^2 u = f_{xx} dx^2 + 2f_{xy} dxdy + f_{yy} dy^2$$

d. **Differential Operator** $d \cdot$, $d = \dfrac{\partial}{\partial x} dx + \dfrac{\partial}{\partial y} dy$ follows the

rule: $du = \left(\dfrac{\partial}{\partial x} dx + \dfrac{\partial}{\partial y} dy \right) u = \dfrac{\partial u}{\partial x} dx + \dfrac{\partial u}{\partial y} dy$.

The **power of operator** d is a binomial, such as:

$$d^2 = \left(\dfrac{\partial}{\partial x} dx + \dfrac{\partial}{\partial y} dy \right)^2 = \dfrac{\partial^2}{\partial x^2} dx^2 + 2 \dfrac{\partial^2}{\partial x \partial y} dxdy + \dfrac{\partial^2}{\partial y^2} dy^2$$

n-th differential is $d^n u = \left(\dfrac{\partial}{\partial x} dx + \dfrac{\partial}{\partial y} dy \right)^n u$

5.4   Taylor's Formula

If function $u = f(x_1, \ldots, x_k)$ belongs to class C^{n+1} in the neighborhood of point M_0, then there exists point N in the interval between N and M_0, such that holds

$$f(M) - f(M_0) = du(M_0) + \dfrac{1}{2!} d^2 u(M_0) + \ldots$$

$$+ \dfrac{1}{n!} d^n u(M_0) + \dfrac{1}{(n+1)!} d^{n+1} u(N)$$

when $dx_i = \Delta x_i = x_i - x_i^0$, $i = 1, 2, \ldots, k$.

Explicit formula for a 2-variable function:

$$f(x,y) = f(x_0, y_0) + [f_x(x_0, y_0)(x - x_0) + f_y(x_0, y_0)(y - y_0)] +$$

$$+ \dfrac{1}{2!} [f_{xx}(x_0, y_0)(x - x_0)^2 + 2f_{xy}(x_0, y_0)(x - x_0)(y - y_0) +$$

$$+ f_{yy}(x_0, y_0)(y - y_0)^2] + \dfrac{1}{(n+1)!} \left(\dfrac{\partial}{\partial x}(x - x_0) + \dfrac{\partial}{\partial y}(y - y_0) \right)^{n+1} f(N)$$

Chapter 6
Chain Rule

a. If function $u = f(x,y,z)$ has partial continuous derivative f_z, f_y, f_x and $x = x(t), y = y(t), z = z(t)$ are derivable, then derivative $\dfrac{du}{dt}$ exists, and there holds:

$$\frac{du}{dt} = f_x \cdot \frac{dx}{dt} + f_y \cdot \frac{dy}{dt} + f_z \cdot \frac{dz}{dt}$$

b. If function $u = f(x,y,z)$ is differentiable, and functions $x = x(t,v)$, $y = y(t,v)$, and $z = z(t,v)$ are also differentiable, then composite function $u = u(t,v)$ is partially derivable by t and v, and there holds:

$$\frac{\partial u}{\partial t} = \frac{\partial f}{\partial x}\cdot\frac{\partial x}{\partial t} + \frac{\partial f}{\partial y}\cdot\frac{\partial y}{\partial t} + \frac{\partial f}{\partial z}\cdot\frac{\partial z}{\partial t} \ , \ \frac{\partial u}{\partial v} = \frac{\partial f}{\partial x}\cdot\frac{\partial x}{\partial v} + \frac{\partial f}{\partial y}\cdot\frac{\partial y}{\partial v} + \frac{\partial f}{\partial z}\cdot\frac{\partial z}{\partial v}.$$

c. **Chain Rule** - general case

If functions $x_i = \varphi_i(t_1, t_2, \ldots, t_m)$, $(i = 1, 2, \ldots, k)$ are differentiable in the neighborhood of $N_0 = (t_1^0, t_2^0, \ldots, t_m^0)$, and function $u = f(x_1, x_2, \ldots, x_k)$ is differentiable in the neighborhood of $M_0(x_1^0, x_2^0, \ldots, x_k^0)$ when $x_i^0 = \varphi_i(N_0)$, then the function composition

$$u = f(\varphi_1(t_1, \ldots, t_m), \varphi_2(t_1, \ldots, t_m), \ldots, \varphi_k(t_1, \ldots, t_m))$$

is differentiable at N_0, and

$$\frac{\partial u}{\partial t_j} = \sum_{i=1}^{k} \frac{\partial u}{\partial x_i}(M_0) \cdot \frac{\partial x_i}{\partial t_j}(N_0) \ , \ j=1,2,\ldots,m$$

Chapter 7
Directional Derivative

a. A directional derivative of function $u = f(x,y,z)$ in the direction of unit vector $\mathbf{a} = (\cos\alpha, \cos\beta, \cos\gamma)$ at point M_0 is $\lim\limits_{M\to M_0} \dfrac{f(M)-f(M_0)}{|\overrightarrow{M_0 M}|}$, when M tends to M_0 along the ray originating from M_0 in the direction of $\mathbf{a}$.

It is denoted as $D_{\mathbf{a}} f(M_0)$, or $\dfrac{\partial}{\partial \mathbf{a}} f(M_0)$.

b. If function $f(x,y,z)$ belongs to class C^1 in the neighborhood of M_0, then

$$D_{\mathbf{a}} f(M_0) = f_x(M_0)\cos\alpha + f_y(M_0)\cos\beta + f_z(M_0)\cos\gamma$$

c. A directional derivative describes the rate of change of the function at a certain point in a certain direction.

d. The maximum rate of change of function $f(M)$ on M_0 is in the direction of gradient vector (see IX.7):

$$\mathbf{g}(M_0) = f_x(M_0)\mathbf{i} + f_y(M_0)\mathbf{j} + f_z(M_0)\mathbf{k}$$

e. The maximum value of directional derivative is in the direction of $\mathbf{g}(M_0)$ and equals

$$D_{\mathbf{g}} f(M_0) = |\mathbf{g}(M_0)| = \sqrt{f_x^2 + f_y^2 + f_z^2}$$

Chapter 8
Implicit Function

a. Function $y = f(x_1, x_2, ..., x_k)$ is called an **implicit function** if it is given as a solution of equation $F(x_1, x_2, ..., x_k, y) = 0$.

b. **Theorem**: If function $F(x_1, x_2, ..., x_k, y)$ is defined in the neighborhood of point $N_0(x_1^0, x_2^0, ..., x_k^0, y_0)$ and holds:

1. $F(x_1^0, x_2^0, ..., x_x^0, y_0) = 0$

2. $F(x_1, x_2, ..., x_k, y)$ belongs to class C^1 in the neighborhood of N_0.

3. $F_y(N_0) \neq 0$

   Then, there exists a neighborhood of N_0 where there exists unique function $y = f(x_1, x_2, ..., x_k)$ which holds $F(x_1, ..., x_k, f(x_1, x_2, ..., x_k)) = 0$ and has the following properties:

   a) $y_0 = f(x_1^0, ..., x_k^0)$

   b) $y = f(x_1, x_2, ..., x_k)$ is continuous at $M_0(x_1^0, ..., x_k^0)$

   c) $y = f(x_1, x_2, ..., x_k)$ is partially derivative at M_0, and

   $$\frac{\partial f}{\partial x_i}(M_0) = -\frac{F_{x_i}(N_0)}{F_y(N_0)} \ , \ i = 1, 2, ..., k$$

c. The theorem applied for a two-variable function: if function $F(x,y)$ holds the conditions mentioned in b.1-3, in the neighborhood of $M_0(x_0,y_0)$, then there exists a neighborhood of M_0 where unique function $y=f(x)$ is defined, such that $F(x,f(x))=0$, and $f(x)$ is continuous and derivable function, the derivative of which is

$$f'(x_0) = -\frac{\Gamma_x(x_0,y_0)}{F_y(x_0,y_0)}$$

Chapter 9
Inverse Functions System

a. Given a system of k functions of k variables:

$$\begin{cases} y_1 = y_1(x_1,x_2,...,x_k) \\ y_2 = y_2(x_1,x_2,...,x_k) \\ \vdots \\ y_k = y_k(x_1,x_2,...,x_k) \end{cases} \quad (*)$$

The determinant

$$J(M) = \frac{D(y_1,y_2,...,y_k)}{D(x_1,x_2,...,x_k)} = \begin{vmatrix} \dfrac{\partial y_1}{\partial x_1} & \dfrac{\partial y_1}{\partial x_2} & \cdots & \dfrac{\partial y_1}{\partial x_k} \\ \dfrac{\partial y_2}{\partial x_1} & \dfrac{\partial y_2}{\partial x_2} & \cdots & \dfrac{\partial y_2}{\partial x_k} \\ \vdots & \vdots & & \vdots \\ \dfrac{\partial y_k}{\partial x_1} & \dfrac{\partial y_k}{\partial x_2} & \cdots & \dfrac{\partial y_k}{\partial x_k} \end{vmatrix}$$

is called the **Jacobian** of (*) system at $M(x_1,x_2,...,x_k)$.

b. If all functions of (*) system belong to class C^1 of the neighborhood of $M_0(x_1^0, x_2^0, \ldots, x_k^0)$ and $J(M_0) \neq 0$, then there exists a neighborhood of $N_0(y_1^0, y_2^0, \ldots, y_k^0)$ when, $y_i(M_0) = y_i^0$, $1 \leq i \leq k$ where there is defined a unique system of functions

$$
\begin{cases}
x_1 = x_1(y_1, y_2, \ldots, y_k) \\
x_2 = x_2(y_1, y_2, \ldots, y_k) \\
\vdots \\
x_k = x_k(y_1, y_2, \ldots, y_k)
\end{cases}
\qquad (**)
$$

inverse to functions y_i of (*), and belonging to class C^1 in the neighborhood of N_0.

c. If $J^* = \dfrac{D(x_1, x_2, \ldots, x_k)}{D(y_1, y_2, \ldots, y_k)}$ is a Jacobian of system (**), then $J \cdot J^* = 1$.

Chapter 10
Applications

a. The equation of the **Tangent line to a plane curve** is expressed by the equation $F(x, y) = 0$ at point $M_0(x_0, y_0)$ is

$$
F_x(M_0)(x - x_0) + F_y(M_0)(y - y_0) = 0
$$

b. The equation of the **Normal to a surface** given by the equation $F(x, y, z) = 0$ at point $M_0(x_0, y_0, z_0)$ is

$$
\frac{x - x_0}{F_x(M_0)} = \frac{y - y_0}{F_y(M_0)} = \frac{z - z_0}{F_z(M_0)}
$$

c. The equation of the **Tangent plane to surface** $F(x,y,z) = 0$ at point $M_0(x_0, y_0, z_0)$ is

$$(x-x_0)F_x(M_0) + (y-y_0)F_y(M_0) + (z-z_0)F_z(M_0) = 0$$

d. **Tangent line to the intersection line of the surfaces** $F(x,y,z) = 0$ and $G(x,y,z) = 0$ at common point $M_0(x_0, y_0, z_0)$. Vector

$$\mathbf{a} = \left(F_x(M_0), F_y(M_0), F_z(M_0)\right) \times \left(G_x(M_0), G_y(M_0), G_z(M_0)\right) =$$

$$= (a_1, a_2, a_3)$$

is in the direction of the tangent line.

Tangent line equation is $\dfrac{x-x_0}{a_1} = \dfrac{y-y_0}{a_2} = \dfrac{z-z_0}{a_3}$.

Chapter 11
Extrema of Multivariable Functions

11.1 Critical Points

a. Function $u = f(x_1, x_2, ..., x_k)$ has a **local maximum** (or **minimum**) on M_0, if there exists a neighborhood of M_0 such that for all M of that neighborhood there holds

$$f(M) \leq f(M_0) \ , \ [f(M) \geq f(M_0)]$$

b. M_0 which is a local maximum or minimum point is called a **point of local extremum** of $f(M)$.

c. **Necessary Condition for the Existence of Extremum**: if function $u = f(x_1, x_2, ..., x_k)$ has an extremum on M_0, and, in addition, it has first-order partial derivatives at that point, then all these derivatives equal zero on M_0.

d. M_0 is a **critical point** of $f(M)$ if all partial derivatives of $f(M)$ equal zero at M_0.

e. All critical points and points at which at least one partial derivative does not exist are **suspected extrema points**.

11.2   Types of Critical Points

If $M_0(x_1^0, x_2^0, ..., x_k^0)$ is a critical point of function $u = f(x_1, x_2, ..., x_k)$ belonging to class C^2 (it has continuous partial derivatives up to second order), and the quadratic form (see next paragraph).

$$d^2 u(M_0) = \left(\frac{\partial}{\partial x_1} dx_1 + ... + \frac{\partial}{\partial x_k} dx_k \right)^2 f(M_0)$$

of variables $dx_1, dx_2, ..., dx_k$ is

1. Positive, then function $u = f(M)$ has local minimum on M_0.

2. Negative, then function $u = f(M)$ has local maximum on M_0.

3. Mixed, then function $u = f(M)$ has no extremum on M_0.

11.3   Analysis of Quadratic Forms

The matrix of quadratic form $d^2 f(M_0)$ (see XI.12) is

$$A = \begin{pmatrix} f_{x_1 x_1} & f_{x_1 x_2} & \cdots & f_{x_1 x_k} \\ f_{x_2 x_1} & f_{x_2 x_2} & \cdots & f_{x_2 x_k} \\ \vdots & \vdots & & \vdots \\ f_{x_k x_1} & f_{x_k x_2} & \cdots & f_{x_k x_k} \end{pmatrix}$$

and

$$A_1 = f_{x_1 x_1} \ , \ A_2 = \begin{vmatrix} f_{x_1 x_1} & f_{x_1 x_2} \\ f_{x_2 x_1} & f_{x_2 x_2} \end{vmatrix} \ , \ A_3 = \begin{vmatrix} f_{x_1 x_1} & f_{x_1 x_2} & f_{x_1 x_3} \\ f_{x_2 x_1} & f_{x_2 x_2} & f_{x_2 x_3} \\ f_{x_3 x_1} & f_{x_3 x_2} & f_{x_3 x_3} \end{vmatrix} \ , \ ... , \ A_k = |A|$$

are k **first minors of matrix A**.

Sylvester Theorem: The form $d^2 f(M_0)$ is positive if, and only if, all first minors of matrix A are positive. Quadratic form $d^2 f(M_0)$ is negative if, and only if, the sign of its first minors change alternately, the first sign being negative.

11.4 Extremum of a Two-Variable Function

Function $u = f(x, y)$ has a local extremum on $M_0(x_0, y_0)$ in which $f_x(M_0) = 0$, $f_y(M_0) = 0$, and

$\Delta = f_{xx}(M_0) \cdot f_{yy}(M_0) - [f_{xy}(M_0)]^2 > 0$. If $f_{xx}(M_0) > 0$, then it is the local minimum, and if $f_{xx}(M_0) < 0$, then it is the local maximum. If $\Delta < 0$, function $u = f(M)$ has no extremum on M_0. In such a case, M_0 is a **saddle** point.

Chapter 12
Extrema with Constraints

Let $u = f(x_1, x_2, ..., x_k)$ be a function with two constraints:

$$G(x_1, ..., x_k) = 0, \ F(x_1, ..., x_k) = 0 \ (*)$$

a. **Definition**: Function $u = f(M)$ has a maximum (or minimum) under the conditions of (*) on $M_0(x_1^0, ..., x_k^0)$ if $F(M_0) = 0$, $G(M_0) = 0$ and there exists a neighborhood of

M_0 such that for all point $M(x_1,...,x_k)$ of that neighborhood which is under the constraints, that is, $F(M) = G(M) = 0$, there holds:

$$f(M) \le f(M_0), \ [f(M) \ge f(M_0)]$$

b. **Lagrange Multiplier Method** of finding critical points with (*) constraints:

Construct a **Lagrange function** $\psi = f + \lambda_1 F + \lambda_2 G$

where λ_1, λ_2 are **Lagrange multipliers**.

c. **A Necessary Condition for the Existence of Extremum**: To find critical points, one must solve a system of $k+2$ equations where

$$\psi_{x_i} = 0 \ , \ i = 1, 2, ..., k \ , \ F = 0 \ , \ G = 0$$

With $k+2$ unknowns which are $x_1,...,x_k, \lambda_1, \lambda_2$ to its solutions are include all the points where its derivatives do not exist.

d. **Absolute Extremum of a Function Continuous Above Bounded and Closed Domain** D: According to Weierstrass theorem, a continuous function has its maximum and minimum value on D.

The function may reach these values of critical points within domain D or on the boundary of D, or on singular points on the boundary of D.

Therefore, to find the maximum and minimum values of the function on a bounded and closed domain, one must find all critical points, (both interior and boundary points, including singular boundary points), and, by comparing the values of the function on these points, absolute maxima and minima in domain D can be found.

VIII: Integral Calculus of Multivariable Functions

Chapter 1
Parameter-Dependent Integral

1.1   Definition

Let $f(x,y)$ be function defined on rectangular

$R = \{(x,y) : 0 \le x \le b, c \le y \le d\}$

a. If $f(x,y)$ is integrable by x in interval $[a,b]$, for every fixed y of $[c,d]$, then function $\varphi(y) = \int\limits_a^b f(x,y)\,dx$ is **integral dependent of the** y **parameter**.

b. If $f(x,y)$ is integrable by y on $[c,d]$ for every $x \in [a,b]$, then function $\psi(x) = \int\limits_a^b f(x,y)\,dy$ is **integral dependent of the** x **parameter**.

1.2   Properties of Parameter-Dependent Integral

a. If function $f(x,y)$ is continuous on rectangle R then $\varphi(y)$ is continuous on $[c,d]$ and $\psi(x)$ is continuous on $[a,b]$.

b. **Leibniz Rule**

1. If $f(x,y)$ and $f_y(x,y)$ are continuous on R , then

$$\frac{d\varphi}{dy} = \frac{d}{dy}\int_a^b f(x,y)\,dx = \int_a^b f_y(x,y)\,dx$$

2. If $f(x,y)$ and $f_x(x,y)$ are continuous on R , then

$$\frac{d\psi}{dx} = \frac{d}{dx}\int_c^d f(x,y)\,dy = \int_c^d f_x(x,y)\,dy$$

c. If $f(x,y)$ is continuous on R , then

$$\int_c^d\left[\int_a^b f(x,y)\,dx\right]dy = \int_a^b\left[\int_c^d f(x,y)\,dy\right]dx$$

d. If $f(x,y)$ is continuous on R and functions $\alpha(y)$, $\beta(y)$ are continuous on $[c,d]$, then $\varphi(y) = \displaystyle\int_{\alpha(y)}^{\beta(y)} f(x,y)\,dx$ is continuous on $[c,d]$.

e. If $f(x,y)$ and $f_y(x,y)$ are continuous on R, and functions $\alpha(y)$, $\beta(y)$ are derivable, then function $\varphi(y)$ is derivable, and

$$\varphi'(y) = \int_{\alpha(y)}^{\beta(y)} f_y(x,y)\,dx + \beta'(y)f\big(\beta(y),y\big) - \alpha'(y)f\big(\alpha(y),y\big)$$

Chapter 2
Double Integral

2.1   Definition of Double Integral

Let $f(x,y)$ be a function defined above region D. Let region D be divided by lines parallel to coordinate lines, into elementary sections, resulting in a rectangular net covering D. The area of inner rectangles Δ_{ij} is $\Delta_{ij} = (x_i - x_{i-1}) \cdot (y_i - y_{i-1}) = \Delta x_i \cdot \Delta y_j$. Inside every rectangle, we choose a point $M_{ij}(u_i, v_j)$ and express a sum by all inner rectangles:

$$\sigma = \sum_{i,j} f(u_i, v_j) \Delta x_i \Delta y_j$$

This is the **integral sum** of function $f(x,y)$ corresponding to the given partition and dependent on the choice of points (u_i, v_j).

a. Function $f(x,y)$ is **integrable (according to Riemann)** above domain D if there exists a finite limit of its integral sum when the maximum diagonal of rectangles Δ_{ij} tends to zero, when the limit is independent both of manner of partitioning and of the choice of points (u_i, v_j).

This limit is the **double integral** of function $f(x,y)$ over D. It is denoted $\iint\limits_{D} f(M)\,d\sigma$ or $\iint\limits_{D} f(x,y)\,dxdy$.

b. Let L be a curve in D and Δ'_{ij} be rectangles covering its area. L is called **zero area curve** if $\lim\limits_{\max \text{diag}\Delta'_{ij} \to 0} \sum\limits_{ij} \Delta'_{ij} = 0$.

c. Function $f(x,y)$ is **piecewise continuous** in D if it discontinuous at no more than a finite number of zero area curves.

d. If $f(x,y)$ is continuous over D then it is integrable.

e. If $f(x,y)$ is integrable over D it is bounded in D.

f. If $f(x,y)$ is bounded and piecewise continuous in D, then it is integrable.

2.2   Properties of Double Integral

a. The area of domain D: $S(D) = \iint\limits_{D} dxdy$.

b. If functions $f(x,y)$ and $g(x,y)$ are integrable in D, then:

1. For every real a and b, function $af(x,y) + bg(x,y)$ is integrable, and

$$\iint\limits_{D} [af(M) \pm bg(M)]d\sigma = a\iint\limits_{D} f(M)d\sigma \pm b\iint\limits_{D} g(M)d\sigma$$

2. Function $f(x,y) \cdot g(x,y)$ is integrable over D.

3. If, in addition, $f(x,y) \le g(x,y)$ on every point of D, then:

$$\iint\limits_{D} f(M)d\sigma \le \iint\limits_{D} g(M)d\sigma$$

c. If $f(x,y)$ is integrable over D, then $|f(x,y)|$ is integrable in the same domain, and

$$\left| \iint_D f(M) d\sigma \right| \le \iint_D | f(M)| d\sigma$$

The inverse is incorrect. That is, from the integrability of $|f(x,y)|$ does not necessarily follow the integrability of $f(x,y)$ in the same domain D.

d. If $f(x,y)$ is integrable in D, $M = \sup_D f(x,y)$, $m = \inf_D f(x,y)$, then $m \cdot S(D) \le \iint_D f(x,y) dx dy \le M \cdot S(D)$, when $S(D)$ is the area of region D.

If, in addition, D is connected and $f(x,y)$ is continuous in D, then there exists a point (x_0, y_0) in D such that $\iint_D f(x,y) dx dy = f(x_0, y_0) \cdot S(D)$.

f. If function $f(x,y)$ is integrable over D, and region D is divided, by curve L of zero area, into connected regions D_1 and D_2 with no inner common points, then $f(x,y)$ is integrable over D_1 and D_2 and holds:

$$\iint_D f(M) d\sigma = \iint_{D_1} f(M) d\sigma + \iint_{D_2} f(M) d\sigma$$

Chapter 3
Double Integral Calculation

a. If the domain is rectangle $R = \{(x,y) : a \le x \le b, c \le y \le d\}$, then:

$$\iint\limits_{R} f(x,y)\,dxdy = \int\limits_{a}^{b} dx \int\limits_{c}^{d} f(x,y)\,dy = \int\limits_{c}^{d} dy \int\limits_{a}^{b} f(x,y)\,dx$$

b. Let function $f(x,y)$ be integrable in domain D_1 of the following properties:

   1. D_1 is bounded and closed.

   2. Any straight line parallel to y-axis intersects the boundary of the region at no more than 2 points $y_1(x)$ and $y_2(x)$, $\left(y_2(x) > y_1(x)\right)$.

   3. Domain D_1 is bounded left and right between straight lines $x = a$ and $x = b$, respectively.
   Then $\iint\limits_{D_1} f(x,y)\,dxdy = \int\limits_{a}^{b} dx \int\limits_{y_1(x)}^{y_2(x)} f(x,y)\,dy$

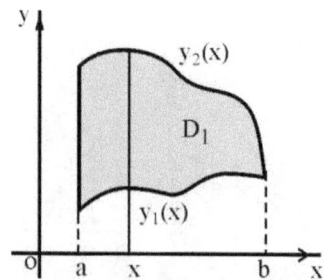

c. If $f(x,y)$ is integrable over bounded and closed region D_2 of the following properties:

1. D_2 is bounded and closed. It is bounded above and below between the straight lines $y = d$ and $y = c$ respectively.

2. Any straight line parallel to the x-axis intersects the boundary of the region at no more than 2 points $x_1(y)$ and $x_2(y)$, $(x_2(y) \geq x_1(y))$.

Then $\iint\limits_{D_2} f(x,y)\,dxdy = \int\limits_c^d dy \int\limits_{x_1(y)}^{x_2(y)} f(x,y)\,dx$

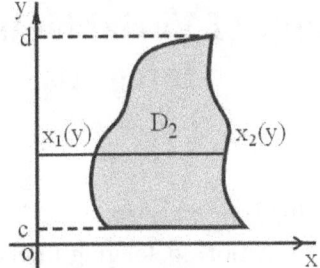

d. If region D is more complex, it is divided into a finite number of domains of the shape of D_1 and D_2.

Example: Illustrated area is divided into 3 regions by a straight line parallel to y-axis.

All regions D_3, D_2, D_1 hold the condition specified in paragraph b, and therefore, the integral can be found using the formula, and:

$$\iint\limits_D f\,d\sigma = \iint\limits_{D_1} f\,d\sigma + \iint\limits_{D_2} f\,d\sigma + \iint\limits_{D_3} f\,d\sigma$$

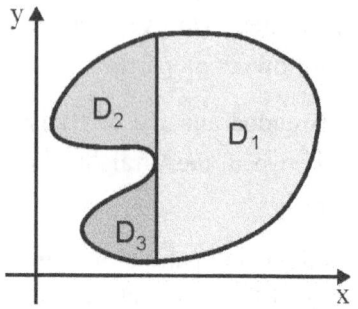

Chapter 4
Change of Variables in Double Integrals

a. Let $x = x(u,v)$, $y = y(u,v)$, $(u,v) \in \Delta$ be a system of continuous functions belonging to class C^1 and mapping domain Δ to domain D with one-to-one correspondence, and the Jacobian (see V.9).

$$J = \frac{D(x,y)}{D(u,v)} = \begin{vmatrix} x_u & x_v \\ y_u & y_v \end{vmatrix}$$

is different from zero at Δ, then

$$\iint\limits_{D} f(x,y)\,dxdy = \iint\limits_{\Delta} f\big(x(u,v),y(u,v)\big)\left| \frac{D(x,y)}{D(u,v)} \right| dudv$$

b. **Transformation to Polar Coordinates**

$$x = \rho\cos t,\ y = \rho\sin t,\ (\rho,t) \in \Delta,\ J = \frac{D(x,y)}{D(\rho,t)} = \rho$$

$$\iint\limits_{D} f(x,y)\,dxdy = \iint\limits_{\Delta} f(\rho\cos t,\rho\sin t)\rho\,d\rho\,dt$$

Chapter 5
Geometric and Physical Applications

a. **Area of Domain** D: $\iint\limits_{D} dxdy = S(D)$

b. **The volume of a cylindrical solid** bounded by a cylindrical surface the directrix of which is parallel to z-axis and bounded above by surface $z_1 = f_1(x,y)$ and below by $z_2 = f_2(x,y)$ is $V = \iint\limits_{D}[f_1(x,y) - f_2(x,y)]dxdy$

when D is the projection of the body on plane x,y.

c. **Mass of planar body** D with planar density $\rho(x,y)$

$$m = \iint\limits_{D} \rho(x,y)dxdy$$

d. **The static moment of inertia** of planar region D, with planar density $\rho(x,y)$

   1. Relative to y-axis $M_y = \iint\limits_{D} x\rho(x,y)dxdy$

   2. Relative to x-axis $M_x = \iint\limits_{D} y\rho(x,y)dxdy$

Chapter 6
Triple Integrals

6.1 Definition

Let function $f(x,y,z)$ be defined above solid (V) in three-dimensional space of volume V. Let us divide (V) into elementary solids $(\Delta v_1),(\Delta v_2),...,(\Delta v_n)$, of volumes $\Delta v_1, \Delta v_2,...,\Delta v_n$, respectively. Let's choose, in every (Δv_i), point $M_i(u_i, t_i, w_i)$ and construct integral sum

$$\sum_i f(u_i, t_i, w_i)\Delta v_i$$

If there exists $\lim \sum_i f(u_i, t_i, w_i)\Delta v_i$, when the maximum diameter of solids (Δv_i) tends to zero and the limit is independent of the partition of (V) and the choice of points M_i, then it is called triple integral of $f(x,y,z)$ over (V) and is denoted as either

$$\iiint\limits_{(V)} f(x,y,z)dxdydz \text{ or } \iiint\limits_{(V)} f(x,y,z)dv \text{ or } \iiint\limits_{(V)} f(M)dv$$

6.2 Properties of Triple Integrals

The properties of triple integrals are similar to those of double integrals (see chap.2.2). Let us just mention the following:

a. **Volume of solid** (V) is $V = \iiint\limits_{(V)} dxdydz$.

b. If function $f(x,y,z)$ is integrable over (V), and holds $m \le f(x,y,z) \le M$ for every $(x,y,z) \in (V)$, then

$$m \cdot V \le \iiint\limits_{(V)} f(x,y,z)\,dxdydz \le M \cdot V$$

c. If function $f(x,y,z)$ is integrable over (V), then $|f(x,y,z)|$ is integrable in the same domain, and

$$\left| \iiint\limits_{(V)} f(x,y,z)\,dv \right| \le \iiint\limits_{(V)} |f(x,y,z)|\,dv$$

6.3   Triple Integrals Calculation

a. If function $f(x,y,z)$ is integrable over rectangular parallelepiped $T = \{(x,y,z) : a \le x \le b, c \le y \le d, e \le z \le f\}$,

then $\iiint\limits_{T} f(x,y,z)\,dv = \int\limits_{a}^{b} dx \int\limits_{c}^{d} dy \int\limits_{e}^{f} f(x,y,z)\,dz$

b. If function $f(x,y,z)$ is integrable above cylindrical body (V) bounded between two surfaces $z = z_1(x,y)$, $z = z_2(x,y)$ above region D on plane (x,y) (Figure 1), then

$$\iiint\limits_{(V)} f(x,y,z)\,dxdydz = \iint\limits_{D} dxdy \int\limits_{z_1(x,y)}^{z_2(x,y)} f(x,y,z)\,dz$$

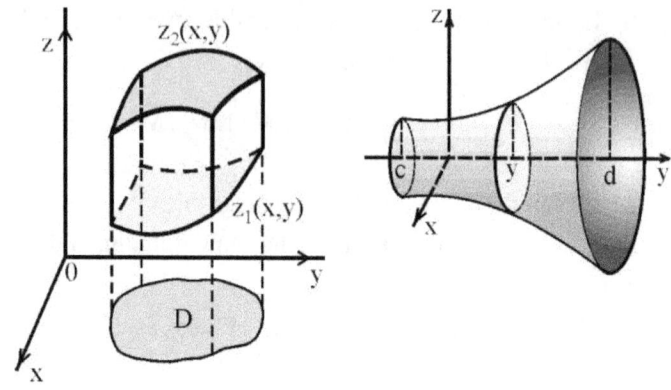

Figure 1   Figure 2

c. If body (V) is between planes $y = c$ and $y = d$, and for every constant y of $[c,d]$ the plane parallel to these planes carves area P_y from body (V) (Figure 2), then

$$\iiint\limits_{(V)} f(x,y,z)\,dv = \int\limits_c^d dy \iint\limits_{P_y} f(x,y,z)\,dzdx$$
.

6.4   Change of Variables in Triple Integrals

a. If the system of functions $x = x(u,t,w)$, $y = y(u,t,w)$, $z = z(u,t,w)$, and $(u,t,w) \in (V')$, continuous and with continuous partial derivatives, maps (V') to body (V) with one-to-one correspondence and the Jacobian

$$\frac{D(x,y,z)}{D(u,t,w)} = \begin{vmatrix} x_u & x_t & x_w \\ y_u & y_t & y_w \\ z_u & z_t & z_w \end{vmatrix}$$

is different from zero, then

$$\iiint\limits_{(V)} f(x,y,z)\,dv = \iiint\limits_{(V')} f\big(x(u,t,w),y(u,t,w),z(u,t,w)\big)$$

$$\left|\frac{D(x,y,z)}{D(u,t,w)}\right| dudtdw$$

b. In cylindrical coordinates $x = \rho\cos t$, $y = \rho\sin t$, $z = z$, the Jacobian is

$$\frac{D(x,y,z)}{D(\rho,\varphi,z)} = \begin{vmatrix} \cos t & -\rho\sin t & 0 \\ \sin t & \rho\cos t & 0 \\ 0 & 0 & 1 \end{vmatrix} = \rho$$

c. In spherical coordinates: $\begin{cases} x = \rho\cos\varphi \cdot \sin\theta \\ y = \rho\sin\varphi \cdot \sin\theta, \\ z = \rho\cos\theta \end{cases}$

the Jacobian is

$$\frac{D(x,y,z)}{D(\rho,\varphi,\theta)} = \begin{vmatrix} \cos\varphi \cdot \sin\theta & -\rho\sin\varphi \cdot \sin\theta & \rho\cos\varphi \cdot \cos\theta \\ \sin\varphi \cdot \sin\theta & \rho\cos\varphi \cdot \sin\theta & \rho\sin\varphi \cdot \cos\theta \\ \cos\theta & 0 & -\rho\sin\theta \end{vmatrix} = -\rho^2 \sin\theta$$

6.5   Applications of Triple Integrals

a. **Volume of the body**: $V = \iiint\limits_{(V)} dxdydz$.

b. **Mass of a body** consisting of a material with specific gravity $\rho(x,y,z)$ is $m = \iiint\limits_{(V)} \rho(x,y,z)dxdydz$.

c. **Static moments** of inertia M_{xy}, M_{yz}, M_{zx} in relation to planes $z = 0$, $x = 0$, and $y = 0$, respectively, is

$$M_{yz} = \iiint\limits_{(V)} x\rho(x,y,z)dv , \quad M_{xy} = \iiint\limits_{(V)} z\rho(x,y,z)dv ,$$

$$M_{xz} = \iiint\limits_{(V)} y\rho(x,y,z)dv$$

d. **Center of mass** (x_0, y_0, z_0) of body (V):

$$x_0 = \frac{M_{yz}}{m}, \quad y_0 = \frac{M_{xz}}{m}, \quad z_0 = \frac{M_{xy}}{m}$$

when M_{xy}, M_{xz}, M_{yz} are static moment calculated using the formulas mentioned in paragraph c, and m is the mass of the body calculated by formula mentioned in paragraph b.

IX. Vector Analysis

Chapter 1
Vector Function, Hodograph

a. **Vector** $r(t) = x(t)\mathbf{i} + y(t)\mathbf{j} + z(t)\mathbf{k}$, $\alpha \le t \le \beta$ when $x(t)$, $y(t)$, and $z(t)$ are real functions with variable t, is called **vector function**.

b. For every constant $t = t_0$, vector function $r(t)$ defines constant vector $r(t_0) = x(t_0)\mathbf{i} + y(t_0)\mathbf{j} + z(t_0)\mathbf{k}$.

c. All operations in vector functions, such as addition, subtraction, multiplication by scalar, inner product and cross product are defined in a similar way (III.4).

d. The locus of the ends of vectors $r(t)$ when $\alpha \le t \le \beta$, which start in the origin, is the **graph** or **path** of vector function $r(t)$ and is called **hodograph**.

e. A graph of vector function can also be represented in a parametric form: $x = x(t)$, $y = y(t)$, $z = z(t)$.

f. **Example**: The graph of vector function

$$r(t) = 2\cos t\,\mathbf{i} + 2\sin t\,\mathbf{j} + t\mathbf{k} \ , \ -\infty < t < \infty$$

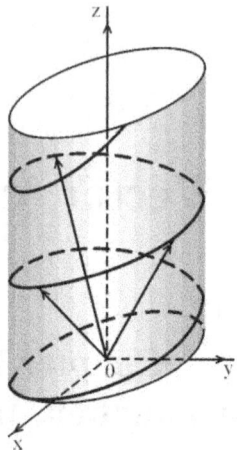

is a curve the parametric equation of which is

$$x = 2\cos t \ , \ y = 2\sin t \ , \ z = t \ , \ -\infty < t < \infty$$

Since for every constant t $x^2 + y^2 = 4$, the end of the vector $\mathbf{r}(t)$ moves along a spiral situated on the side of a circular cylinder (see illustration).

Chapter 2
Limits and Continuity of Vector Functions

a. Constant vector $\mathbf{a} = (a_1, a_2, a_3)$ is the limit of vector function $\mathbf{r}(t) = (x(t), y(t), z(t))$ when $t \to t_0$ if for every $\varepsilon > 0$ there exists $\delta > 0$ such that for every t holding $0 < |t - t_0| < \delta$ there holds $|\mathbf{r}(t) - \mathbf{a}| < \varepsilon$. Written $\lim_{t \to t_0} \mathbf{r}(t) = \mathbf{a}$.

b. $\lim_{t \to t_0} \mathbf{r}(t) = \mathbf{a}$ if and only if $\lim_{t \to t_0} x(t) = a_1$, $\lim_{t \to t_0} y(t) = a_2$, $\lim_{t \to t_0} z(t) = a_3$.

c. The limit of the sum of two vector functions is equal to the sum of their limits.

d. The limit of an inner (cross) product of two vector functions is equal to the inner (cross) product of their limits.

e. Function $\mathbf{r}(t)$ is continuous at point t_0 if it is defined in the neighborhood of t_0, and $\lim_{t \to t_0} \mathbf{r}(t) = \mathbf{r}(t_0)$.

f. The sum, inner product and cross product of continuous vector functions is a continuous vector function.

Chapter 3
Derivative of Vector Function

$$r(t) = x(t)\mathbf{i} + y(t)\mathbf{j} + z(t)\mathbf{k}$$

a. **Definition**: Let us add Δt to t_0, then:

$\Delta \mathbf{r} = \mathbf{r}(t_0 + \Delta t) - \mathbf{r}(t_0) =$

$= [x(t_0 + \Delta t) - x(t_0)]\mathbf{i} + [y(t_0 + \Delta t) - y(t_0)]\mathbf{j} + [z(t_0 + \Delta t) - z(t_0)]\mathbf{k}$

If the limit $\mathbf{r}'(t_0) = \lim\limits_{\Delta t \to 0} \dfrac{\Delta \mathbf{r}}{\Delta t}$ exists, then it is the **derivative**

of vector function $\mathbf{r}(t)$ on t_0.

b. $\mathbf{r}'(t_0) = x'(t_0)\mathbf{i} + y'(t_0)\mathbf{j} + z'(t_0)\mathbf{k}$

c. Geometric description of $\mathbf{r}'(t_0)$: In this illustration, t_0 corresponds to point $M_0\left(x(t_0), y(t_0), z(t_0)\right)$ on the graph of $\mathbf{r}(t)$, $\Delta \mathbf{r} = \overrightarrow{M_0 M}$. If $\Delta t \to 0$, then $M \to M_0$.

The limit of vector $\overrightarrow{M_0 M}$ when $M \to M_0$, is a tangent line to graph of $\mathbf{r}(t)$, at point M_0. Therefore, vector $\mathbf{r}'(t_0)$ is in the direction of the tangent line to the graph of $\mathbf{r}(t)$, at t_0.

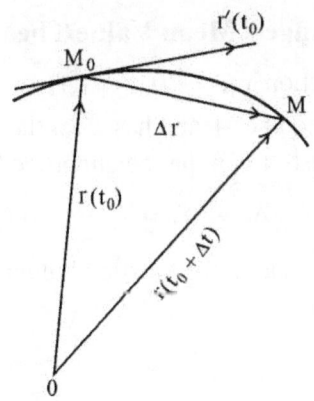

Chapter 4
Properties of the Derivatives

4.1   Derivative Rules

If $\mathbf{r}(t)$, $\mathbf{v}(t)$ are derivable vector functions, $\mathbf{c}$ a constant vector and $f(t)$ a scalar function, then:

1. $\mathbf{c}' = \mathbf{0}$

2. $\left(f(t)\mathbf{r}(t)\right)' = f'(t)\mathbf{r}(t) + f(t)\mathbf{r}'(t)$

3. $\left(\mathbf{r}(t) \pm \mathbf{v}(t)\right)' = \mathbf{r}'(t) \pm \mathbf{v}'(t)$

4. $\left(\mathbf{r}(t) \cdot \mathbf{v}(t)\right)' = \mathbf{r}'(t) \cdot \mathbf{v}(t) + \mathbf{r}(t) \cdot \mathbf{v}'(t)$

5. $\left(\mathbf{r}(t) \times \mathbf{v}(t)\right)' = \mathbf{r}'(t) \times \mathbf{v}(t) + \mathbf{r}(t) \times \mathbf{v}'(t)$

6. If $\mathbf{r}(t)$ is a vector function of constant length $\left(|\mathbf{r}(t)| = \text{const}\right)$, then $\mathbf{r}(t)$ is perpendicular to $\mathbf{r}'(t)$.

4.2   Lagrange's Mean Value Theorem

If vector function $\mathbf{r}(t) = x(t)\mathbf{i} + y(t)\mathbf{j} + z(t)\mathbf{k}$ is continuous in the neighborhood of t and has a continuous derivative on t, then there exists $t*$ of that neighborhood where:

$$\mathbf{r}(t+\Delta t) - \mathbf{r}(t) = \mathbf{r}'(t*) \cdot \Delta t + \vec{\varepsilon}(t, \Delta t) \cdot \Delta t$$

when $\vec{\varepsilon}(t, \Delta t)$ is a vector function holding $\lim_{\Delta t \to 0} \vec{\varepsilon}(t, \Delta t) = \mathbf{0}$.

Chapter 5
Arc Length

a. The length of a path presented in its vector form

$$\mathbf{r}(t) = x(t)\mathbf{i} + y(t)\mathbf{j} + z(t)\mathbf{k} \quad , \quad \alpha \le t \le \beta$$

is $\ell = \int_{\alpha}^{\beta} |\mathbf{r}'(t)| \, dt = \int_{\alpha}^{\beta} \sqrt{\left(x'(t)\right)^2 + \left(y'(t)\right)^2 + \left(z'(t)\right)^2} \, dt$

b. If $y = f(x)$, $a \le x \le b$, then the arc length is $\ell = \int_{a}^{b} \sqrt{1 + \left(f'(x)\right)^2} \, dx$.

c. If the curve is presented in polar coordinates, that is $\rho = \rho(\varphi)$, $\varphi_0 \le \varphi \le \varphi_1$, then arc length is

$$\ell = \int_{\varphi_0}^{\varphi_1} \sqrt{[\rho(\varphi)]^2 + [\rho'(\varphi)]^2} \, d\varphi$$

Chapter 6
Vector Functions of Two Variables

a. Vector

$$\mathbf{V}(t,u) = x(t,u)\mathbf{i} + y(t,u)\mathbf{j} + z(t,u)\mathbf{k} \quad , \quad (t,u) \in \Delta \quad (*)$$

is a **vector function** dependent of variables t and u and defined in domain Δ.

b. The graph of a two-variable vector function is a surface in a three-dimensional space, presented in a parametric form

$$x = x(t,u) \ , \ y = y(t,u) \ , \ z = z(t,u) \ , \ (t,u) \in \Delta$$

vise versa, any surface in space can be presented in a vector form (*).

Example:

1. Using spherical coordinates, the vector representation of sphere $x^2 + y^2 + z^2 = R^2$ is

$$\mathbf{V}(\rho,\theta) = R\cos\varphi\cos\theta\mathbf{i} + R\sin\varphi\cos\theta\mathbf{j} + R\sin\theta$$

$$0 \le \varphi \le 2\pi \quad , \quad -\frac{\pi}{2} \le \theta \le \frac{\pi}{2}$$

2. Surface $z = f(x,y)$, $(x,y) \in D$ presented in a vector form:

$$\mathbf{V}(x,y) = x\mathbf{i} + y\mathbf{j} + f(x,y)\mathbf{k}, \ (x,y) \in D$$

c. **Normal to a Surface**

1. If $M_0(t_0,u_0)$ be a point on surface $\mathbf{V}(t,u)$, then vector $\mathbf{V}_t(t_0,u_0)$ is in the direction of the tangent line to curve $\mathbf{V}(t,u_0)$ and belongs to the tangent plane to the surface at M_0.

   Vector $\mathbf{V}_u(t_0,u_0)$ is in the direction of the tangent line to curve $\mathbf{V}(t_0,u)$, which is on the tangent plane to the surface at the same point, (t_0,u_0).

   Normal $\mathbf{N}$, which is perpendicular to the tangent plane, that is, to the plane of vectors $\mathbf{V}_u(t_0,u_0)$ and $\mathbf{V}_t(t_0,u_0)$, is

   $$\mathbf{N} = \mathbf{V}_t(t_0,u_0) \times \mathbf{V}_u(t_0,u_0) = \begin{vmatrix} \mathbf{i} & \mathbf{j} & \mathbf{k} \\ x_t & y_t & z_t \\ x_u & y_u & z_u \end{vmatrix} (M_0)$$

   The unit vector of the normal is $\mathbf{n} = \dfrac{\mathbf{V}_t \times \mathbf{V}_u}{|\mathbf{V}_t \times \mathbf{V}_u|}$

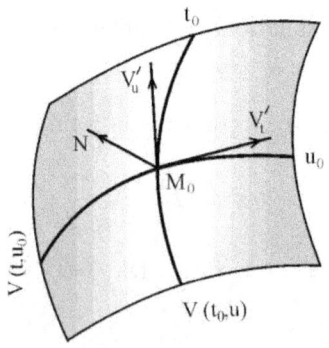

2. Normal to a surface, given in its explicit form of $z = f(x, y)$ is

$$N = -f_x(x_0, y_0)i - f_y(x_0, y_0)j + k$$

3. Normal to a surface presented in the form of $F(x, y, z) = 0$ is

$$N = F_x(x_0, y_0, z_0)i + F_y(x_0, y_0, z_0)j + F_z(x_0, y_0, z_0)k$$

Chapter 7
Scalar Field Gradient

a. The function $f(M)$ given in every point of body V, defined a **scalar field**.

b. Through every point $M_0 \in V$, just one level surface $f(M) = f(M_0)$ passes (see VII.2.2).

c. The change rate of a differentiable scalar field at M_0 in the direction of vector $a = (\cos\alpha, \cos\beta, \cos\gamma)$ is directional derivative (see VII.7).

$$D_a f(M_0) = f_x(M_0)\cos\alpha + f_y(M_0)\cos\beta + f_z(M_0)\cos\gamma$$

d. Vector $f_x i + f_y j + f_z k$ is called **gradient** of function $u = f(x, y, z)$, and is denoted $\operatorname{grad} f$ or ∇f when operator ∇ (**Nabla**) is

$$\nabla = \frac{\partial}{\partial x}i + \frac{\partial}{\partial y}j + \frac{\partial}{\partial z}k$$

operating on function u, following the rule

$$\nabla u = \frac{\partial u}{\partial x} \mathbf{i} + \frac{\partial u}{\partial y} \mathbf{j} + \frac{\partial u}{\partial z} \mathbf{k}$$

e. A gradient of scalar field at M_0 is in the direction of the normal to the level surface passing through M_0 (or a level curve, if the field is planar).

f. Maximum change rate of scalar fiend $f(M)$ is in the direction of vector ∇f.

g. The maximum value of the directional derivative of function $u = f(x, y, z)$ on M_0 is

$$\max D_a f(M_0) = |\nabla f(M_0)| = \sqrt{f_x^2(M_0) + f_y^2(M_0) + f_z^2(M_0)}$$

Chapter 8
Vector Field

a. If in every point $M(x, y, z)$ on body V vector

$\mathbf{F}(x, y, z) = P(x, y, z)\mathbf{i} + Q(x, y, z)\mathbf{j} + R(x, y, z)\mathbf{k}$ is defined, then

$\mathbf{F}(x, y, z)$ is called a **vector field** above V.

b. **Planar vector** field

$\mathbf{F}(x, y) = P(x, y)\mathbf{i} + Q(x, y)\mathbf{j}$, $(x, y) \in D$.

c. For every scalar field $u = f(x, y, z)$ there is a corresponding vector field which is gradient u:

$$f_x(M)\mathbf{i} + f_y(M)\mathbf{j} + f_z(M)\mathbf{k} = \nabla u$$

d. The expression $\dfrac{\partial P}{\partial x}(x,y,z)+\dfrac{\partial Q}{\partial y}(x,y,z)+\dfrac{\partial R}{\partial z}(x,y,z)$ is

called the **divergent of vector field** $\mathbf{F}$, and is denoted $\mathrm{div}\mathbf{F}$.

$$\mathrm{div}\mathbf{F} = \nabla \cdot \mathbf{F}$$

e. Vector

$$\left(\frac{\partial R}{\partial y}-\frac{\partial Q}{\partial z}\right)\mathbf{i}+\left(\frac{\partial P}{\partial z}-\frac{\partial R}{\partial x}\right)\mathbf{j}+\left(\frac{\partial Q}{\partial x}-\frac{\partial P}{\partial y}\right)\mathbf{k}$$

is called the **rotor of vector field**

$$\mathbf{F} = P\mathbf{i}+Q\mathbf{j}+R\mathbf{k}$$

and is denoted $\mathrm{rot}\mathbf{F}$. Using operator ∇, we write

$$\mathrm{rot}\mathbf{F} = \nabla \times \mathbf{F} = \begin{vmatrix} \mathbf{i} & \mathbf{j} & \mathbf{k} \\ \dfrac{\partial}{\partial x} & \dfrac{\partial}{\partial y} & \dfrac{\partial}{\partial z} \\ P & Q & R \end{vmatrix}$$

f.

1. $\mathrm{div}(\mathrm{rot}\mathbf{F}) = 0$

2. $\mathrm{div}(\mathbf{F} \pm \mathbf{G}) = \mathrm{div}\,\mathbf{F} \pm \mathrm{div}\,\mathbf{G}$

3. $\mathrm{div}(f \cdot \mathbf{F}) = f\,\mathrm{div}\mathbf{F} + \mathbf{F} \cdot \mathrm{gard}\,f$

4. $\mathrm{rot}(\mathbf{F} \pm \mathbf{G}) = \mathrm{rot}\,\mathbf{F} \pm \mathrm{rot}\,\mathbf{G}$

5. $\mathrm{rot}(f \cdot \mathbf{F}) = f\,\mathrm{rot}\mathbf{F} + (\mathrm{grad}\,f) \times \mathbf{F}$

6. $\mathrm{rot}(\mathrm{gard}\,f) = 0$

7. $\mathrm{div}(\mathrm{grad}\,f) = \Delta f = f''_{xx} + f''_{yy} + f''_{zz}$

8. $\mathrm{div}(\mathbf{F} \times \mathbf{G}) = \mathbf{G} \cdot \mathrm{rot}\,\mathbf{F} - \mathbf{F} \cdot \mathrm{rot}\,\mathbf{G}$

Chapter 9
Line Integrals

9.1   Definition

Given: Curve $\ell : \mathbf{r}(t) = x(t)\mathbf{i} + y(t)\mathbf{j} + z(t)\mathbf{k}$, $\alpha \le t \le \beta$

And vector field $\mathbf{F} = P(x,y,z)\mathbf{i} + Q(x,y,z)\mathbf{j} + R(x,y,z)\mathbf{k}$

a. Line ℓ is a **smooth curve** if $\mathbf{r}(t), \mathbf{r}'(t)$ are continuous vector functions.

b. The **positive direction** on ℓ is the same as the direction of parameter t increase. If ℓ is a closed line, then we define its equation such that when the parameter increases, its direction on ℓ is such that during the movement along ℓ in this direction, the finite domain is bounded by it remains on the left.

c. **Line integral**, or integral along ℓ is

$$\int_{\ell} P(x,y,z)\,dx + Q(x,y,z)\,dy + R(x,y,z)\,dz = \int_{\ell} \mathbf{F} \cdot \mathbf{r}'\,dt = \int_{\ell} \mathbf{F} \cdot d\mathbf{r}$$

d. If $\mathbf{F} = P(x,y)\mathbf{i} + Q(x,y)\mathbf{j}$ is a planar vector field, and curve ℓ is given in its explicit form, $\ell : \{y = f(x), a \le x \le b\}$, then:

$$\int_{\ell} \mathbf{F} \cdot d\mathbf{r} = \int_{a}^{b} [P\big(x, y(x)\big) + Q\big(x, y(x)\big) \cdot y'(x)]\,dx$$

e. The physical meaning of line integral is **work** done by field $\mathbf{F}$ of moving a particle from point $t = \alpha$ to $t = \beta$ along ℓ .

9.2   Properties of Line Integrals

a. $\int\limits_{\ell} (a\mathbf{F} + b\mathbf{G})\,\mathbf{dr} = a\int\limits_{\ell} \mathbf{F}\cdot\mathbf{dr} + b\int\limits_{\ell} \mathbf{G}\cdot\mathbf{dr}$

b. $|\int\limits_{\ell} \mathbf{F}\cdot\mathbf{dr}| \le \max |\mathbf{F}|\cdot|\ell|$, when $|\ell|$ is the length of ℓ.

c. If curve L consists of a finite number of curves $\ell_1, \ell_2, ..., \ell_k$, then $\int\limits_{L} \mathbf{F}\cdot\mathbf{dr} = \sum_{i=1}^{k} \int\limits_{\ell_i} \mathbf{F}\cdot\mathbf{dr}$.

d. The line integral depends on the direction defined on ℓ. If the same route is passed in the opposite direction, the integral will change its sign: $\int\limits_{AB} \mathbf{F}\cdot\mathbf{dr} = -\int\limits_{BA} \mathbf{F}\cdot\mathbf{dr}$.

9.3   Green's Theorem: Conservative Field in the Plane

a. **The positive direction on the boundary curve** Γ (consisting of a finite number of closed and piecewise smooth lines) of region D is such that if we move along all parts of Γ in this direction, region D will be on the left.

b. **Green's Theorem**: Let D be a connected and open domain with boundary Γ in the positive direction, and let $\mathbf{F}(x,y) = P(x,y)\mathbf{i} + Q(x,y)\mathbf{j}$ vector field of C^1 in D. Then:

$$\int\limits_{\Gamma} P(x,y)\,dx + Q(x,y)\,dy = \iint\limits_{D} \left(\frac{\partial Q}{\partial x} - \frac{\partial P}{\partial y} \right) dxdy$$

Or, in a vector form: $\int\limits_{\Gamma} \mathbf{F}\cdot\mathbf{dr} = \iint\limits_{D} \mathbf{k}\cdot\text{rot}\mathbf{F}\,dxdy$

c. If $\mathbf{F}(x,y) = P(x,y)\,\mathbf{i} + Q(x,y)\,\mathbf{j}$ is a vector field defined above planar domain D, then all the following propositions are equivalent:

1. $\int_{\Gamma} \mathbf{F} \cdot \mathbf{dr} = 0$ for every closed line Γ in D.

2. The integral $\int_{AB} \mathbf{F} \cdot \mathbf{dr}$ is independent of the path connecting points A and B, which is entirely in D.

3. There exists continuous scalar function $u(x,y)$ holding $\nabla u = \mathbf{F}(x,y)$, in other words, the expression $P(x,y)dx + Q(x,y)dy$ is a full differential, $du = Pdx + Qdy$, and there holds

$$\int_{AB} Pdx + Qdy = u(B) - u(A)$$

4. If, in addition, field $\mathbf{F}$ is of class C^1 and domain D is **simply connected**, that is, its boundary consists of a single closed line not intersecting itself, then the last three propositions are equivalent to

$$\frac{\partial P}{\partial y} = \frac{\partial Q}{\partial x}$$

d. The vector field $\mathbf{F}(x,y)$ is a **conservative field** in D if there exists scalar field $u(x,y)$ such that $\nabla u = \mathbf{F}$. Function $u(x,y)$ is called a **potential of field F**.

The function of a potential of a vector field is a unique up to a constant.

e. **Formulas of potential function** $u(x,y)$.

If $\mathbf{F}$ is a conservative field in D and $(x_0, y_0) \in D$, then:

$$u(x,y) = \int_{x_0}^{x} P(t,y)dt + \int_{y_0}^{y} Q(x_0,v)dv + c$$

or
$$u(x,y) = \int_{y_0}^{y} Q(x,v)\,dv + \int_{x_0}^{x} P(t,y_0)\,dt + c$$

f. If D is a bound domain with boundary ∂D and $\mathbf{n} = (\cos\alpha, \cos\beta)$ is a unit normal to D directed outwards and $\Delta u = u_{xx} + u_{yy}$, then:

$$\iint_{D} (v\Delta u - u\Delta v)\,dxdy = \int_{\partial D} \left(v\frac{\partial u}{\partial \mathbf{n}} - u\frac{\partial v}{\partial \mathbf{n}} \right) d\ell$$

Chapter 10
Surface Integral

10.1 Surfaces

$$S : \mathbf{r}(u,v) = x(u,v)\mathbf{i} + y(u,v)\mathbf{j} + z(u,u)\mathbf{k} \ , \ (u,v) \in \Delta$$

a. S is a **smooth surface** if vector function $\mathbf{r}(u,v)$ is of class C^1 region.

b. The **unit normal** to surface S is $\mathbf{n} = \dfrac{\mathbf{r}_u \times \mathbf{r}_v}{|\mathbf{r}_u \times \mathbf{r}_v|}$

c. A smooth surface S is called oriented (two-sided) if there is a normal unit vector n at every point on S, not on the boundary curve of S, such that n is a continuous function of (x, y, z) on S. Later on, each surface must be oriented, have a finite area and a single smooth, closed boundary curve. The sphere, ellipsoid, paraboloid, and hyperboloid are two-sided surfaces.

A nonorientable surface is, for example, a Mobius strip. It is formed by half-twisting a strip of paper, attaching its ends together (see illustration). If we draw a unit normal N from one point of a Mobius strip, we eventually get a normal at the same point, at the opposite direction of the original normal.

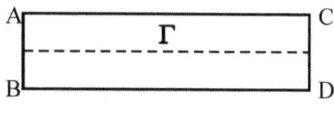

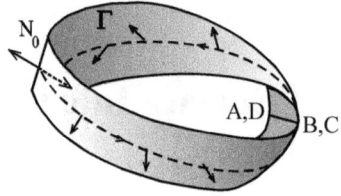

d. A **smooth surface** S is two-sided if the direction of normal can on it can be specified definitely. **Later on, we shall only refer to two-sided surfaces.**

e. **The area of the surface is** $S = \iint\limits_{\Delta} |\mathbf{r}_u \times \mathbf{r}_v| \, dudv$.

f. If surface S is given in the form $z = f(x,y), (x,y) \in D$, then its area is $S = \iint\limits_{D} \sqrt{1 + (f_x)^2 + (f_y)^2} \, dxdy$.

g. **The positive direction** of surface S is the direction of normal $\mathbf{N} = \mathbf{r}_u \times \mathbf{r}_v$. If S is the boundary of a closed body, then the positive direction on S is outwards. If S is an open surface, then the positive direction is towards the Z-axis.

10.2 Surface Integral

Given a vector field

$$\mathbf{F}(x,y,z) = P(x,y,z)\mathbf{i} + Q(x,y,z)\mathbf{j} + R(x,y,z)\mathbf{k}$$

a. $\iint_S \mathbf{F} \cdot \mathbf{n}\, ds$ is a **surface integral** when $\mathbf{n}$ is the unit normal

vector to surface S.

b.   Different ways of writing and calculating surface integral:

$$\iint_S \mathbf{F} \cdot \mathbf{n}\, ds = \iint_S P\,dydz + Q\,dxdz + R\,dxdy =$$

$$\iint_S (P\cos\alpha + Q\cos\beta + R\cos\gamma)\,ds =$$

$$= \iint_\Delta \mathbf{F} \cdot \mathbf{r}_u \times \mathbf{r}_v\, dudv = \iint_\Delta \begin{vmatrix} P & Q & R \\ x_u & y_u & z_u \\ x_v & y_v & z_v \end{vmatrix} dudv$$

when α, β, γ are directional angles of $\mathbf{n}$.

c.   **Gauss Divergent Theorem**

Let V be a body the boundary of which is smooth and closed surface S with an outer normal $\mathbf{n}$. If vector field $\mathbf{F} = P\mathbf{i} + Q\mathbf{j} + R\mathbf{k}$ is of class C^1 in V, then holds

$$\iint_S \mathbf{F} \cdot \mathbf{n}\, ds = \iiint_V \operatorname{div} \mathbf{F}\, dv$$

or

$$\iint_S (P\cos\alpha + Q\cos\beta + Q\cos\gamma)\,ds = \iiint_V \left(\frac{\partial P}{\partial x} + \frac{\partial Q}{\partial y} + \frac{\partial R}{\partial z} \right) dxdydz.$$

The integral $\iint_S \mathbf{F} \cdot \mathbf{n}\, ds$ is called the **flux of vector field $\mathbf{F}$**

through surface S.

d. Stokes Theorem

If, in the neighborhood of two-sided surface S, vector field $\mathbf{F} = P\mathbf{i} + Q\mathbf{j} + R\mathbf{k}$ is of class C^1 and Γ is the boundary of surface S in the positive direction, then

$$\iint_S \nabla \times \mathbf{F} \cdot \mathbf{n} ds = \int_\Gamma \mathbf{F} \cdot \mathbf{dr} \ \text{ or } \ \iint_S \operatorname{rot} \mathbf{F} \cdot \mathbf{n} ds = \int_\Gamma \mathbf{F} \cdot \mathbf{dr}$$

denote

$$\iint_S \operatorname{rot} \mathbf{F} \cdot \mathbf{n} \, ds = \iint_\Delta \operatorname{rot} \mathbf{F} \cdot \mathbf{r}_u \times \mathbf{r}_v \, du dv =$$

$$\iint_\Delta \begin{vmatrix} R_y - Q_z & P_z - R_x & Q_x - P_y \\ x_u & y_u & z_u \\ x_v & y_v & z_v \end{vmatrix} du dv$$

Chapter 11
Conservative Field in General

a. Vector field $\mathbf{F}$ is called **conservative field** if there exists scalar field $u(x, y, z)$ such that $\nabla u = \mathbf{F}$.

b. Vector field $\mathbf{F} = P\mathbf{i} + Q\mathbf{j} + R\mathbf{k}$ of class C^1 above simply connected surface body V (that is, on every closed curve L or S there exists a two-sided surface, of which L is its boundary) is conservative if, and only if, one of the following propositions hold:

1. $\int_L \mathbf{F} \cdot \mathbf{dr} = 0$ for every simple closed line L in V.

2. Integral $\int\limits_{AB} \mathbf{F} \cdot \mathbf{dr}$ is independent of the line connecting points A and B, which is entirely in V.

3. There exists scalar field $u(x,y,z)$ holding $\nabla u = \mathbf{F}(x,y,z)$. In other words, the expression $Pdx + Qdy + Rdz$ is a full differential. It means there exists a function $u(x,y,z)$ such that $du = Pdx + Qdy + Rdz$ and there holds

$$\int\limits_{AB} \mathbf{F} \cdot \mathbf{dr} = u(B) - u(A).$$

4. $\nabla \times \mathbf{F} = \text{rot}\,\mathbf{F} = 0$.

c. If filed $\mathbf{F}$ is conservative in V and (x_0, y_0, z_0) is a point in V, then the potential function can be calculated using one of the following formulas:

$$u(x,y,z) = \int\limits_{x_0}^{x} P(t,y,z)\,dt + \int\limits_{y_0}^{y} Q(x_0,v,z)\,dv + \int\limits_{z_0}^{z} R(x_0,y_0,w)\,dw + c$$

$$u(x,y,z) = \int\limits_{y_0}^{y} Q(x,v,z)\,dv + \int\limits_{x_0}^{x} P(t,y_0,z)\,dt + \int\limits_{z_0}^{z} R(x_0,y_0,w)\,dw + c$$

X. Algebra

Chapter 1
Complex Numbers

1.1 Definition: Algebraic Operations in Complex Numbers

a. An ordered pair of real numbers $z = (x, y)$ is called a **complex number**.

b. $z = x + iy$, when $i^2 = -1$, is the **algebraic form** of a complex number.

   x is called the **real part** of z, and is denoted $x = \operatorname{Re} z$.

   y is called the **imaginary part** of z, and is denoted $y = \operatorname{Im} z$.

c. $\bar{z} = x - iy$ is a **conjugate** to the number $z = x + iy$

d. Let $z_1 = x_1 + y_1 i$ and $z_2 = x_2 + y_2 i$ be two complex numbers:

   1. **Equality**: $z_1 = z_2$ if and only if $x_1 = x_2$ and $y_1 = y_2$

   2. **Addition**: $z_1 + z_2 = (x_1 + x_2) + (y_1 + y_2)i$

   3. **Multiplication**: $z_1 z_2 = (x_1 x_2 - y_1 y_2) + (x_1 y_2 + x_2 y_1)i$

   4. **Powers of** i : $i^2 = -1$, $i^3 = -i$, $i^4 = 1$

5. **Division:** $\dfrac{z_1}{z_2} = \dfrac{x_1 x_2 + y_1 y_2}{x_2^2 + y_2^2} + \dfrac{x_2 y_1 - x_1 y_2}{x_2^2 + y_2^2} i$

6. $\overline{z_1 + z_2} = \overline{z}_1 + \overline{z}_2$, $\overline{z_1 \cdot z_2} = \overline{z}_1 \cdot \overline{z}_2$, $\overline{\left(\dfrac{z_1}{z_2}\right)} = \dfrac{\overline{z}_1}{\overline{z}_2}$

1.2 Geometric Description, Modulus and Argument

a. Any complex number $z = a + ib = (a, b)$ can be described as point in plane x, y or a vector beginning in the origin and ending at point (a, b).

Plane (x, y) is called **complex plane**, when x-axis is **Real Axis** and y-axis is **Imaginary Axis**.

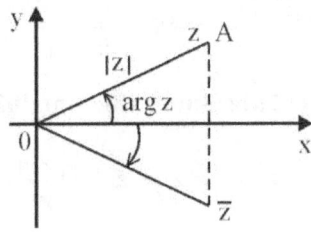

b. In the complex plane, $\overline{z}$ is described by a vector symmetrical to z about x-axis.

c. The length of vector $\overrightarrow{OA}$ s called the **modulus** or **absolute value** of complex number $z = a + ib$ and is denoted $|z| = \sqrt{a^2 + b^2}$.

d. Angle φ between the vector corresponding to complex number z and the positive direction of the real axis is called the **argument** of z and is denoted as $\arg z$. It is measured counterclockwise, in radians.

Notice: For every complex number, the modulus is specified definitely, but an argument has countless values different from each other by an integer product of 2π.

e. $z = r(\cos\varphi + i\sin\varphi)$ is the **polar form** of a complex number. To have its polar form, we should just take an argument value of $0 \leq \varphi < 2\pi$, or, alternately, $-\pi \leq \varphi < \pi$, since an addition of $2\pi n$ for φ does not change the values of $\cos\varphi$ and $\sin\varphi$.

f. The relation between the algebraic form $z = a + bi$ and the polar form $z = r(\cos\varphi + i\sin\varphi)$ of a complex number is:

$$a = r\cos\varphi \ , \ b = r\sin\varphi$$

$$r = |z| = \sqrt{a^2 + b^2} \ , \ \cos\varphi = \frac{a}{\sqrt{a^2 + b^2}} \ , \ \sin\varphi = \frac{b}{\sqrt{a^2 + b^2}}$$

g. **Properties of absolute value (modulus)**

$$|z_1 \cdot z_2| = |z_1| \cdot |z_2| \ , \ \left|\frac{z_1}{z_2}\right| = \frac{|z_1|}{|z_2|}$$

$$\left||z_1| - |z_2|\right| \leq |z_1 + z_2| \leq |z_1| + |z_2|$$

h. **Properties of the argument**

$$\arg(z_1 \cdot z_2) = \arg(z_1) + \arg(z_2)$$

$$\arg\left(\frac{z_1}{z_2}\right) = \arg(z_1) - \arg(z_2)$$

$$\arg\overline{z} = -\arg z$$

1.3   Powers and Roots of $z = r(\cos\varphi + i\sin\varphi)$

a. **De Moivre's Formula**: For every natural n,

$$z^n = r^n(\cos n\varphi + i\sin n\varphi)$$

b. **The n-th root of a complex number** z is a complex number w holding $w^n = z$. It is denoted $\sqrt[n]{z} = w$.

c. For every complex number $z = r(\cos\alpha + i\sin\alpha)$, $r \neq 0$ there exist just n different complex numbers $w_0, w_1, w_2, \ldots, w_{n-1}$ for which $w_k^n = z$, expressed by the formula

$$w_k = \sqrt[n]{r}\left(\cos\frac{\alpha + 2\pi k}{n} + i\sin\frac{\alpha + 2\pi k}{n}\right), \quad k = 0,1,2,\ldots,n-1$$

Chapter 2
Fields

a. A **field** is a set of at least two terms, in which two **closed operations**, addition and multiplication, are defined. That is, for every a and b of the field, $a+b$ and $a \cdot b$, are of the field, and have the following properties:

1. Addition is commutative: $a+b = b+a$.

2. Addition is associative: $a+(b+c) = (a+b)+c$.

3. There exists one element, 0, indifferent to addition. That is, for very element a of $0+a = a$.

4. For each a there is one element, $-a$, called additive inversed, such that $a+(-a) = 0$.

5. Multiplication is commutative: $a \cdot b = b \cdot a$.

6. Multiplication is associative: $(a \cdot b) \cdot c = a \cdot (b \cdot c)$

7. There exists one element, 1, indifferent to multiplication. That is, for every a of $a \cdot 1 = a$.

8. For every $a \neq 0$ there is one element a^{-1}, such that $a \cdot a^{-1} = 1$.

9. Multiplication is distributive with respect to addition: $(a + b) \cdot c = a \cdot c + b \cdot c$.

The terms of a field are called **scalars**.

b. $\mathbb{R}$ is the real numbers field.

c. $\mathbb{C}$ is the complex numbers field.

d. $\mathbb{Q}$ is the rational numbers field.

Chapter 3
Polynomials

3.1 Definition

a. $P_n(x) = a_n x^n + a_{n-1} x^{n-1} + \ldots + a_1 x + a_0 = \sum_{i=0}^{n} a_i x^i$, $a_n \neq 0$ is an n-**th-degree polynomial**. It is denoted $\deg P_n(x) = n$.

b. If $a_i \in \mathbb{R}$, $i = 0, 1, \ldots, n$, it is a **real polynomial**.

c. If $a_i \in \mathbb{C}, i = 0, 1, \ldots, n$, it is a **complex polynomial**.

d. $a_i, i = 0, 1, \ldots, n-1$ are **coefficients of the polynomial**, a_n is the **leading coefficient**.

e. A constant different than zero is a **zero-degree polynomial**.

f. A polynomial with all its coefficients equal to zero is called a **zero polynomial**. A zero polynomial has no degree.

3.2   Sum and Product of Polynomials

Let $P_n(x) = \sum_{i=0}^{n} a_i x^i$ and $Q_m(x) = \sum_{i=0}^{m} b_i x^i$ be two polynomials.

a. Polynomials $P_n(x)$ and $Q_m(x)$ are **equal** if and only if $m = n$ and $a_i = b_i$ for all $i = 0, 1, \ldots, n$.

b. We define the sum of two polynomials as a collection of like terms.

   For example: If $n \geq m$, then $P_n(x) + Q_m(x) = \sum_{i=0}^{n} c_i x^i$ when

   $$c_i = \begin{cases} a_i + b_i & , i = 0,1,2,\ldots,m \\ a_i & , i = m+1, m+2, \ldots, n \end{cases}$$

c. The degree of a sum of polynomials is smaller than or equal to the maximum of their degrees: $\deg[P_n(x) + Q_m(x)] \leq \max[\deg P_n(x), \deg Q_m(x)]$.

d. The product of two polynomials is

   $P_n(x) \cdot Q_m(x) = \sum_{i=0}^{n+m} d_i x^i$, when $d_i = \sum_{k+\ell=i} a_k \cdot b_\ell$

   $i = 0, 1, \ldots, m+n$ (the sum is by every k and ℓ such that $\ell + k = i$).

e. The degree product of polynomials equals to the sum of their degrees:

   $$\deg[P_n(x) \cdot Q_m(x)] = \deg P_n(x) + \deg Q_m(x)$$

3.3   Polynomial Division

a. The division of polynomial $P_n(x)$ by polynomial $D_m(x)$, $m \le n$ is represented the following way:

$$P_n(x) = D_m(x) \cdot Q(x) + R(x) \ \text{ or } \ \frac{P_n(x)}{D_m(x)} = Q(x) + \frac{R(x)}{D_m(x)} \ \ (*)$$

when $Q(x)$ is the **quotient**, $R(x)$ is the **remainder**, and $\deg R(x) < m$.

If $R(x) \equiv 0$, then polynomial $P_n(x)$ is **divided without a remainder** by $D_m(x)$.

b. For every two polynomials $P_n(x)$ and $D_m(x)$ there exists a **unique** pair of polynomials $Q(x)$ and $R(x)$ holding (*) and $\deg R(x) < \deg D_m(x)$ or $R(x) = 0$.

c. For the division of polynomials we use the same algorithm as for the division of real numbers.

3.4   Remainder Theorem

a. Remainder R after the division of polynomial $P(x)$ by $(x - a)$ equals to the value of polynomial at $x = a$, that is $R = P(a)$.

b. Polynomial $P(x)$ is divided by $x - a$ without remainder, if and only if $P(a) = 0$.

c. Number a zeroing the polynomial, $\left(P(a) = 0\right)$, is called **root of the polynomial**.

3.5   Factorization of Polynomials

a. Every polynomial, real or complex, of a degree greater than zero, has at least one complex root.

b. Every complex polynomial, $P_n(x) = \sum_{i=0}^{n} a_i x^i$, $(n > 0)$ has
exactly n roots, not necessarily different, $x_1, x_2, \ldots, x_n$,
and

$$P_n(x) = a_n(x - x_1) \cdot (x - x_2) \cdot \ldots \cdot (x - x_n)$$

c. If polynomial $P_n(x)$ is divisible without remainder by
$(x - x_i)^{k_i}$, $i = 1, 2, \ldots, m$, and indivisible by $(x - x_i)^{k_i + 1}$,
then k_i is called the **multiplicity** of root x_i and x_i is
called **root with multiplicity** k_i .

d. $x = a$ is the root of polynomials $P_n(x)$ with multiplicity
k , $(k < n)$, if, and only if,
$P_n(a) = 0, P'_n(a) = 0, \ldots, P_n^{(k-1)}(a) = 0$, $P^{(k)}(a) \neq 0$.

e. If $x = a + bi$ is the root of polynomial $P_n(x) = \sum_{k=0}^{n} a_k x^k$
with real coefficients, then $\overline{x} = a - bi$ is also a root of it.

3.6   Vieta's Formulas

Vieta's formulas connect the roots and coefficients of
polynomial.

If $x_1, x_2, \ldots, x_n$ are roots of polynomial $\sum_{k=0}^{n} a_k x^k$, then:

$$x_1 + x_2 + \ldots + x_n = -\frac{a_{n-1}}{a_n}$$

$$x_1 x_2 + x_1 x_3 + \ldots + x_{n-1} x_n = \frac{a_{n-2}}{a_n}$$

$$x_1 x_2 x_3 + x_1 x_2 x_4 + \ldots + x_{n-2} x_{n-1} x_n = -\frac{a_{n-3}}{a_n}$$

$$\ldots \qquad \ldots \qquad \ldots \qquad \ldots$$

$$x_1 \cdot x_2 \cdot \ldots \cdot x_n = (-1)^n \cdot \frac{a_0}{a_n}$$

Chapter 4
Vector Spaces $\mathbb{R}^n$ and $\mathbb{C}^n$

4.1   Definition

a. An ordered set of n scalars $(a_1, a_2, \ldots, a_n)$ of field $\mathbb{F}$ is a **vector**.

b. Let $\mathbf{a} = (a_1, a_2, \ldots, a_n)$ be a vector above $\mathbb{F}$ and k a scalar of $\mathbb{F}$. A **vector multiplied by a scalar** is vector $k\mathbf{a} = (ka_1, ka_2, \ldots, ka_n)$.

c. The **sum of vectors** $\mathbf{a}(a_1, a_2, \ldots, a_n)$ and $\mathbf{b} = (b_1, b_2, \ldots, b_n)$ is vector $\mathbf{a} + \mathbf{b} = (a_1 + b_1, a_2 + b_2, \ldots, a_n + b_n)$.

d. $\mathbb{R}^n$ $(\mathbb{C}^n)$ is called **n-dimensional real (complex) vector space**.

e. Non-empty subset W of $\mathbb{F}^n$ is called **subspace** if:

   1. For every pair of vectors $\mathbf{a}$ and $\mathbf{b}$ of W, vector $\mathbf{a} + \mathbf{b}$ is also of W.

2. For every scalar $\alpha \in \mathbb{F}$ and vector $\mathbf{a} \in W$, vector $\alpha \mathbf{a} \in W$.

4.2   Linear Dependency and Independency

a. Vector $\quad \mathbf{u} = \sum_{i=1}^{k} \alpha_i \cdot \mathbf{u}_i = \alpha_1 \mathbf{u}_1 + \alpha_2 \mathbf{u}_2 + \ldots + \alpha_k \mathbf{u}_k$, $\quad$ when

$\alpha_i \in \mathbb{F}$, $\quad \mathbf{u}_i \in \mathbb{F}^n$, $\quad i = 1, 2, \ldots, k$ $\quad$ is called a **linear combination** of k vectors $\mathbf{u}_i$.

b. The set of all linear combinations

$$W = \{\mathbf{u} = \sum_{i=1}^{k} \alpha_i \mathbf{a}_i, \alpha_i \in \mathbb{F}, \mathbf{a}_i \in \mathbb{F}^n\}$$

is called **subspace spanned** by $\mathbf{a}_1, \ldots, \mathbf{a}_k$ or the **span** of vectors $\mathbf{a}_1, \ldots, \mathbf{a}_k$. It is denoted

$$W = \text{span}(\mathbf{a}_1, \mathbf{a}_2, \ldots, \mathbf{a}_k)$$

c. A set of vectors $\mathbf{a}_1, \mathbf{a}_2, \ldots, \mathbf{a}_k \in \mathbb{F}^n$ are **linearly dependent** if there exist scalars $\alpha_1, \alpha_2, \ldots, \alpha_k \in \mathbb{F}$, not all of which are zero, such that

$$\alpha_1 \mathbf{u}_1 + \alpha_2 \mathbf{u}_2 + \ldots + \alpha_k \mathbf{u}_k = 0$$

Otherwise, the set is said to be **linearly independent, or LI.**

d. Vectors $\mathbf{u}$ and $\mathbf{v}$ are linearly dependent if and only if they are proportionate, that is, there exists scalar α such that $\mathbf{u} = \alpha \mathbf{v}$ or $\mathbf{v} = \alpha \mathbf{u}$.

e. If set of vectors $\mathbf{u}_1, \mathbf{u}_2, \ldots, \mathbf{u}_n$ is linearly independent, then every of its subsets is linearly independent.

f. Set of non-zero vectors S (finite or infinite) is **linearly dependent** if and only if it has a finite subset of linearly dependent vectors.

S is **linearly independent** if it has no finite set of linearly dependent vectors.

4.3 Basis and Dimensions of Vector Spaces

a. A set of vectors S is called a **basis** of subspace $V \in \mathbb{F}^n$ if:

1. S is linearly independent.

2. $V = \text{span} S$

b. The vectors

$$\mathbf{e}_1 = (1,0,0,\ldots,0)$$
$$\mathbf{e}_2 = (0,1,0,\ldots,0)$$
$$\mathbf{e}_3 = (0,0,1,\ldots,0)$$
$$\vdots \qquad \vdots \qquad \vdots$$
$$\mathbf{e}_n = (0,0,0,\ldots,1)$$

Are the **natural basis** of $\mathbb{F}^n$.

c. The number of independent vectors in a subspace is not greater than the number of the vectors spanning it.

d. If $\{\mathbf{v}_1,\ldots,\mathbf{v}_s\}$ and $\{\mathbf{u}_1,\ldots,\mathbf{u}_r\}$ are bases of subspace W, then $s = r$.

e. The number of vectors in a basis of subspace W is called the **dimension** of W and is denoted $\dim W$.

A vector space with a dimension n is called an n-**dimensional space**.

f. Any set of less than n vectors in an n-dimensional space V can not span V.

g. Any set of more than n vectors in an n-dimensional space is linearly dependent.

Chapter 5
Matrices

5.1   Definition and Types of Matrices

a. A table of $m \times n$ real (complex) numbers arranged in m rows and n columns

$$A = \begin{pmatrix} a_{11} & a_{12} & \cdots & a_{1n} \\ a_{21} & a_{22} & \cdots & a_{2n} \\ \vdots & \vdots & & \vdots \\ a_{m1} & a_{m2} & \cdots & a_{mn} \end{pmatrix} = (a_{ij})_{i=1,\,j=1}^{m,n}$$

is an $m \times n$ **order real (complex) matrix**.

b. $\mathbb{F}^{m \times n}$ is the set of all $m \times n$ matrices with $\mathbb{F}$ field elements.

c. $\mathbf{a}_i = (a_{i1}, a_{i2}, \ldots, a_{in})$ is the i -**row vector** on A, $\mathbf{a}_i \in \mathbb{F}^{1 \times n}$.

d. $\mathbf{b}_j = \begin{pmatrix} a_{1j} \\ a_{2j} \\ \vdots \\ a_{mj} \end{pmatrix}$ is the j -**column vector** in A, $\mathbf{b}_j \in \mathbb{F}^{m \times 1}$.

e. The **transposed matrix** of A, A^t, results from switching the rows to columns. That is, if $A \in \mathbb{F}^{m \times n}$ then $A^t \in \mathbb{F}^{n \times m}$ and

$$(A^t)_{ij} = a_{ji}$$

f. A matrix where the number of rows equals the number of columns is called n -**th order square matrix**.

g. A matrix in which all of its elements are zeroes is a **zero matrix**.

h. A square matrix $A = (a_{ij})_{i,j=1}^{n}$ is called **diagonal matrix** if for every $a_{ij} = 0, i \neq j$, that is, all elements except those on the main diagonal are zeroes. It denoted $\mathrm{diag}(a_{11}, a_{22}, \ldots, a_{nn})$.

i. A diagonal matrix is **scalar** if all terms of the main diagonal equal each other.

j. An **identity matrix** is a scalar matrix where the elements of the diagonal equal 1. An n-th order identity matrix is denoted I_n.

k. **Upper (lower) triangular matrix** is a square matrix where all elements below (above) the main diagonal are zeroes. In other words, matrix $A \in \mathbb{F}^{n \times n}$ is upper triangular if $a_{ij} = 0$ for every $i > j$, and lower triangular when $a_{ij} = 0$ for every $i < j$.

l. Square matrix $A \in \mathbb{R}^{n \times n}$ is called **symmetric** if $a_{ij} = a_{ji}$ for every i, j, that is, $A^t = A$.

m. If, in matrix A, $a_{ij} = -a_{ji}$, for every i, j that is, $A^t = -A$ then, it is called **anti-symmetric** matrix. In an anti-symmetric matrix, element a_{ii} situated on the main diagonal equal zero.

n. Any square matrix A can be presented as the sum of a symmetric and an anti-symmetric matrix. That is, $A = C + B$, when $C = \frac{1}{2}(A + A^t)$ is a symmetric matrix, and $B = \frac{1}{2}(A - A^t)$, an anti-symmetric one.

o. Complex matrix $\overline{A} = \left(\overline{a_{ij}}\right)_{i,j=1}^{n}$ is a **conjugate transpose** of matrix $A = (a_{ij})_{i,j=1}^{n}$. It means that $\overline{A}$ is obtained by taking conjugate complex of every element of A.

p. Matrix A is called **Hermitian** if $A = \overline{A^t}$.

q. Matrix A is called **anti-Hermitian** if $A = -\overline{A^t}$.

r. Matrix A is called **normal** if $\overline{A^t} \cdot A = A \cdot \overline{A^t}$.

s. Matrix A is called **unitary** if $\overline{A^t} \cdot A = A \cdot \overline{A^t} = I$.

t. A real unitary matrix is called an **orthogonal matrix.**

5.2   Algebraic Operations on Matrices

a. Two **matrices are equal** if they are of the same order and the elements of the same locations are equal.

b. The **product of matrix** $A \in \mathbb{F}^{m \times n}$ **multiplied by scalar** $\alpha \in \mathbb{F}$ is matrix αA resulted from the multiplication of all A elements by α.

c. The **sum** of matrices A and B of $\mathbb{F}^{m \times n}$ is the matrix resulted from summing their corresponding elements. That is: $(A + B)_{ij} = a_{ij} + b_{ij}$, $i = 1, 2, \ldots, m$, $j = 1, \ldots, n$.

d. The **inner product** of vectors $\mathbf{a} = (a_1, a_2, \ldots, a_m)$ and $\mathbf{b} = (b_1, b_2, \ldots, b_m)$ of $\mathbb{F}^m$ is scalar
$\mathbf{a} \cdot \mathbf{b} = a_1 b_1 + a_2 b_2 + \ldots + a_m b_m$.

e. **The multiplication** $A \cdot B$ of matrices $A \in \mathbb{F}^{m \times n}$ and $B \in \mathbb{F}^{n \times k}$ is $m \times k$ order matrix C, when its c_{ij} element is the inner multiplication of i-row vector of matrix A by j-column vector of matrix B:

$$c_{ij} = a_{i1}b_{1j} + a_{i2}b_{2j} + \ldots + a_{in}b_{nj} = \sum_{\ell=1}^{n} a_{i\ell}b_{\ell j} \ , \ i = 1,\ldots,m \ ; \ j = 1,2,\ldots,k$$

Example: $\begin{pmatrix} 2 & 3 \\ 4 & 5 \end{pmatrix} \cdot \begin{pmatrix} a & b & c \\ d & e & f \end{pmatrix} = \begin{pmatrix} 2a+3d & 2b+3e & 2c+3f \\ 4a+5d & 4b+5e & 4c+5f \end{pmatrix}$

f. If $A \in \mathbb{F}^{m \times n}$; $B, D \in \mathbb{F}^{n \times k}$; $C \in \mathbb{F}^{k \times r}$, then:

1. $A(BC) = (AB)C$

2. $A(B+D) = AB + AD$

3. $(B+D) \cdot C = BC + DC$

4. $A \cdot I_n = I_m \cdot A = A$

5. $(AB)^t = B^t \cdot A^t$

g. **The trace** of square matrix $A \in \mathbb{F}^{n \times n}$ is the sum of elements on its main diagonal $trA = \sum_{i=1}^{n} a_{ii}$.

1. $tr(A+B) = trA + trB$

2. $tr(AB) = tr(BA)$

5.3   Row Space and Columns Space

a. The **row space** of matrix $A \in \mathbb{F}^{m \times n}$ is the span of m rows on A.

b. The **column space** of $A \in \mathbb{F}^{n \times m}$ is the span of its columns.

c. The row space dimension of A is equal to its column space dimension and is called the **rank** of matrix A, rankA .

5.4   Elementary Row Operations, Staircase Matrix

a. The **elementary row operations** of matrix $A = (a_{ij})_{i=1, j=1}^{n \quad ,m}$ are:

1. Interchanging row i with row k : $(a_i) \leftrightarrow (a_k)$.

2. Multiplying row i by a non-zero scalar, $c \neq 0$, $c(a_i) \to (a_i)$.

3. Adding c times row k to row i : $c(a_k) + (a_i) \to (a_i)$.

b. Matrix A is a **staircase matrix** if:

1. All rows of zeroes, if it has such, are at the bottom.

2. Every non-zero first element is right of the non-zero first element of the last row.

c. Number of non-zero rows in the echelon form matrix resulted from matrix A by elementary row operation equals to the **rank** of matrix A.

5.5   Invertible Matrix

a. Square matrix $A \in \mathbb{F}^{n \times n}$ is **invertible** if there exists a matrix $B \in \mathbb{F}^{n \times n}$ such that

$$AB = BA = I$$

where I is identity matrix.

Matrix B is called **inverse** of A and is denoted A^{-1}.

b. If $A, B \in \mathbb{F}^{n \times n}$ are invertible matrices, then:

1. A^t in invertible and $(A^t)^{-1} = (A^{-1})^t$

2. Multiplication $A \cdot B$ is invertible, and $(A \cdot B)^{-1} = B^{-1} \cdot A^{-1}$.

c. To find A^{-1}, we construct matrix $(A|I)$, and using elementary row operations, get from matrix A to identity matrix I, getting, A^{-1} instead of I.

Example: Find the A^{-1} to matrix $A = \begin{pmatrix} 1 & 0 & 2 \\ 2 & -1 & 3 \\ 4 & 1 & 8 \end{pmatrix}$

$$(A|I) = \begin{pmatrix} 1 & 0 & 2 & | & 1 & 0 & 0 \\ 2 & -1 & 3 & | & 0 & 1 & 0 \\ 4 & 1 & 8 & | & 0 & 0 & 1 \end{pmatrix} \xrightarrow[(-4)(a_1)+(a_3)\to(a_3)]{(-2)(a_1)+(a_2)\to(a_2)}$$

$$\begin{pmatrix} 1 & 0 & 2 & | & 1 & 0 & 0 \\ 0 & -1 & -1 & | & -2 & 1 & 0 \\ 0 & 1 & 0 & | & -4 & 0 & 1 \end{pmatrix} \xrightarrow{(-1)(a_2)\leftrightarrow(a_3)}$$

$$\begin{pmatrix} 1 & 0 & 2 & | & 1 & 0 & 0 \\ 0 & 1 & 0 & | & -4 & 0 & 1 \\ 0 & 1 & 1 & | & 2 & -1 & 0 \end{pmatrix} \xrightarrow{-(a_2)+(a_3)\to a_3}$$

$$\begin{pmatrix} 1 & 0 & 2 & | & 1 & 0 & 0 \\ 0 & 1 & 0 & | & -4 & 0 & 1 \\ 0 & 0 & 1 & | & 6 & -1 & -1 \end{pmatrix} \xrightarrow{(a_1)-2(a_3)\to(a_1)} \begin{pmatrix} 1 & 0 & 0 & | & -11 & 2 & 2 \\ 0 & 1 & 0 & | & -4 & 0 & 1 \\ 0 & 0 & 1 & | & 6 & -1 & -1 \end{pmatrix} =$$

$$= (I|A^{-1})$$

The result is $A^{-1} = \begin{pmatrix} -11 & 2 & 2 \\ -4 & 0 & 1 \\ 6 & -1 & -1 \end{pmatrix}$

5.6   One-sided Invertibility

a. Matrix A of $m \times n$ order of is **left invertible** of there exists matrix B of $n \times m$ order, such that $BA = I_n$. Matrix B is called the **left inverse** of A.

b. Matrix $A \in \mathbb{F}^{m \times n}$ is **right invertible** if there exists matrix $C \in \mathbb{F}^{n \times m}$, such that $AC = I_m$. C is called the **right inverse** of A.

Chapter 6
Determinants

6.1   Second-Order and Third-Order Determinants

a. $a_{11} \cdot a_{22} - a_{12} \cdot a_{21}$ is called second-order determinant of matrix $A = \begin{pmatrix} a_{11} & a_{12} \\ a_{21} & a_{22} \end{pmatrix}$ and is denoted

$$\det A = \begin{vmatrix} a_{11} & a_{12} \\ a_{21} & a_{22} \end{vmatrix} = |A| = a_{11}a_{22} - a_{12}a_{21}$$

b. The third order determinant of the matrix is the number:

$$\det A = \begin{vmatrix} a_{11} & a_{12} & a_{13} \\ a_{21} & a_{22} & a_{23} \\ a_{31} & a_{32} & a_{33} \end{vmatrix} = a_{11} \begin{vmatrix} a_{22} & a_{23} \\ a_{32} & a_{33} \end{vmatrix} - a_{12} \begin{vmatrix} a_{21} & a_{23} \\ a_{31} & a_{33} \end{vmatrix} + a_{13} \begin{vmatrix} a_{21} & a_{22} \\ a_{31} & a_{32} \end{vmatrix}$$

6.2   Permutations and n-th Order Determinant

a. The arrangement $(i_1, i_2, \ldots, i_n)$ of n integers $(1, 2, \ldots, n)$ is called **permutation**.

b. All n! permutations of $(1, 2, \ldots, n)$ is denoted S_n.

c. Let σ be a permutation in S_n. A **disorder** in σ is a pair of integers k, m $(k, m \le n)$, when $m > k$, but m precedes k in σ.

d. σ is an **even permutation**, if the number of disorders is even. On the other hand, it is an odd permutation if the number of disorders is odd. The sign of σ is denoted by $\operatorname{sgn}\sigma$.

$$\operatorname{sgn}\sigma = \begin{cases} 1, \text{if } \sigma \text{ is even} \\ -1, \text{if } \sigma \text{ is odd} \end{cases}$$

e. Switching the places of two elements in a permutation changes the evenness (or oddness).

f. For every square matrix $A = (a_{ij})_{ij=1}^{n}$ the determinant is:

$$|A| = \det A = \sum_{S_n} \operatorname{sgn}(i_1,\ldots,i_n) a_{1i_1} a_{2i_2} \ldots a_{ni_n}$$

when the sum includes all $n!$ permutations $(i_1, i_2, \ldots, i_n)$ of numbers $(1, 2, \ldots, n)$.

In other words: The determinant of matrix A is the sum of $n!$ elements in the form of $a_{1i_1} a_{2i_2} \ldots a_{ni_n}$ every one of which has a **single representative of each row and each column** with the corresponding sign.

g. Scalar $|M_{ij}|$, called the **cofactor** a_{ij} of matrix A is an $(n-1)$-th order determinant of a matrix resulted from eliminating row i and column j of matrix A .

h. Expansion by row r : The determinant of n-th order square matrix A equals to $|A| = \sum_{j=1}^{n} (-1)^{r+j} a_{rj} |M_{rj}|$.

6.3 Properties of Determinants

a. A determinant of upper or lower triangular or diagonal matrix equals to the product of diagonal elements.

b. If all elements in a row (column) of a matrix are zeroes, then its determinant equals zero.

c. If n-th order matrices A, B, C are only different by the elements of row k, that is, $a_{ij} = b_{ij} = c_{ij}$, $i \neq k$, $j = 1, 2, \ldots, n$, and $c_{kj} = a_{kj} + b_{kj}$, then $\det A + \det B = \det C$.

d. If matrix B is obtained from matrix A, by multiplying all the elements of just one row by an α, then $|B| = \alpha |A|$.

e. The determinant of a matrix which has two equal rows, equals zero.

f. If two rows in a matrix are proportionate, then its determinant equals zero.

g. A determinant does not change if all the elements of one row are added corresponding elements of another row multiplied by a non-zero constant.

h. A determinant of a multiplication of matrices is equal to the multiplication of their determinants:

$$\det(A \cdot B) = \det A \cdot \det B$$

i. $\det A = \det A^t$.

j. If $A \in \mathbb{C}^{n \times n}$, then $\det \overline{A} = \overline{\det A}$.

6.4   Inversion of Matrices and Determinants

a. Matrix A s invertible, if and only if $\det A \neq 0$.

b. If A is invertible, then $|A^{-1}| = \dfrac{1}{|A|}$.

c. If matrix $A = (a_{ij})_{i,j=1}^n$ is invertible, then

$$A^{-1} = \frac{1}{|A|} \cdot \begin{pmatrix} A_{11} & A_{21} & \cdots & A_{n1} \\ A_{12} & A_{22} & & A_{n2} \\ \vdots & \vdots & & \vdots \\ A_{1n} & A_{2n} & \cdots & A_{nn} \end{pmatrix}$$

when $A_{ij} = (-1)^{i+j} |M_{ij}|$ is the **algebraic cofactor** of element a_{ij} and $|M_{ij}|$ is a cofactor corresponding to a_{ij} (see 6.2.g).

Chapter 7
System of Linear Equations

7.1   Definition and Solution

a. A system of m linear equations with n unknowns:

$$\begin{cases} a_{11}x_1 + a_{12}x_2 + \cdots + a_{1n}x_n = b_1 \\ a_{21}x_1 + a_{22}x_2 + \cdots + a_{2n}x_n = b_2 \\ \vdots \qquad \vdots \qquad \qquad \vdots \qquad \vdots \\ a_{m1}x_1 + a_{m2}x_2 + \cdots + a_{mn}x_n = b_m \end{cases} \quad (*)$$

when $x_1, x_2, ..., x_n$ are unknowns, $a_{ij}; 1 \le i \le m, 1 \le j \le n$ are their coefficients, with the first index i indicating which equation coefficient a_{ij} is, and the second index j, indicating that a_{ij} is the coefficient of x_j.

b. $b_1, b_2, ..., b_m$ are the **free elements**.

c. (*) is a homogeneous system if all its free coefficients equal zero.

d. The **coefficient matrix** of system (*) is:

$$A = \begin{pmatrix} a_{11} & a_{12} & \cdots & a_{1n} \\ a_{21} & a_{22} & \cdots & a_{2n} \\ \vdots & \vdots & & \vdots \\ a_{m1} & a_{m2} & \cdots & a_{mn} \end{pmatrix}$$

e. $Ax = b$ is a vector form of system (*) when

$$\mathbf{b} = \begin{pmatrix} b_1 \\ b_2 \\ \vdots \\ b_m \end{pmatrix} \quad , \quad \mathbf{x} = \begin{pmatrix} x_1 \\ x_2 \\ \vdots \\ x_n \end{pmatrix}$$

f. The **extended matrix** of system (*) is

$$(A \mid \mathbf{b}) = \begin{pmatrix} a_{11} & \cdots & a_{1n} & b_1 \\ a_{21} & \cdots & a_{2n} & b_2 \\ \vdots & & \vdots & \vdots \\ a_{m1} & \cdots & a_{mn} & b_m \end{pmatrix}.$$

g. The n numbers $x_1, x_2, \ldots, x_n$ are the solution of equation system (*) if, when substituted in **each equation**, we get an identity.

$\mathbf{x} = (x_1, x_2, \ldots, x_n)$ is the **solution vector**.

Homogeneous system always has a solution $x_1 = \ldots = x_n = 0$.

$\mathbf{x} = (0, 0, \ldots, 0)$ is called **trivial solution**.

h. The equation system solution does not change if:

1. Both sides of an equation are multiplied by a non-zero number.

2. A multiple of one equation is added to another equation.

3. The positions of two equations are swapped.

i. A system of m equations with n unknowns, $Ax = b$, has:

1. A **unique solution** if and only if the rank of matrix A is equal to the rank of matrix $(A \mid b)$ and equal to n.

2. An **infinite number of solutions** occurs when $\text{rank}A = \text{rank}(A \mid b) < n$. In such a case, the system has $n - \text{rank}A$ **degrees of freedom**.

3. There is no solution, when $\text{rank}A \neq \text{rank}(A \mid b)$.

j. A system of n linear equations with n unknowns has a unique solution, if and only if $\text{rank}A = n$.

k. Homogeneous system $Ax = 0$ has at least one solution, such as the trivial solution.

l. If, in a homogeneous system, number of unknowns n is greater than number of equations m, then the system has an infinite number of solutions, and its general solution has at least $n - m$ **degrees of freedom**.

m. **Cramer theorem**: a system $Ax = b$ with n equations and n unknowns has a unique solution if and only if $|A| \neq 0$. The unique solution is expressed by:

$$x_i = \frac{D_i}{|A|} \ , \ i = 1,2,\ldots,n$$

when D_i is a determinant of the matrix obtained from A when column i is substituted with b.

7.2   Null Space

a. The set of solutions of system $Ax = b$, when $A \in \mathbb{F}^{m \times n}$ is a subspace of $\mathbb{F}^n$ if and only if $b = 0$.

b. The subspace of $Ax = 0$ is referred to as the **solution space** or the **null space** of A.

c. The dimension of the null space of A is $n - \text{rank}A$.

d. If y is a solution of $Ax = b$, and x^0, $x^1,\ldots, x^k$ are basis of the null space of A, then the general solution of $Ax = b$ can be expressed as

$$x = y + c_1 x^1 + c_2 x^2 + ... + c_k x^k$$

Chapter 8
General Vector Spaces

8.1   Definition and Examples

a. **Vector space** V over field $\mathbb{F}$ is a nonempty set of vectors, where two operations are defined:

1. **Addition**: for every two vectors $a, b \in V$, vector $a + b \in V$.

2. **Scalar multiplication**: for every $a \in V$ and $\alpha \in \mathbb{F}$, vector $\alpha a \in V$. These operations follow the following axioms:

For every $a, b, c \in V$ and every $\alpha, \beta \in \mathbb{F}$ there holds:

1. $a + b = b + a$

2. $(a + b) + c = a + (b + c)$

3. In V, there exists a unique **zero vector** holding $a + 0 = a$.

4. In V, there exists a unique term $-a$, called the **opposite vector of a**, which holds $a + (-a) = 0$.

5. $1 \cdot a = a$ when 1 is the **identity element** of field $\mathbb{F}$.

6. $(\alpha\beta)a = \alpha(\beta a)$

7. $\alpha(a + b) = \alpha a + \alpha b$

8. $(\alpha + \beta)a = \alpha a + \beta a$

b. Examples of vector spaces:

1. $\mathbb{R}^n$ is a real vector space.

2. $\mathbb{C}^n$ is a complex vector space.

3. Set of all matrices $\mathbb{F}^{m \times n}$ with elements of field $\mathbb{F}$.

4. $\mathbb{F}[x]$, set of all polynomials with coefficients of field $\mathbb{F}$.

5. $\mathbb{F}_n[x]$, set of all polynomials with a degree smaller than n, including zero polynomial, the coefficients of which are of $\mathbb{F}$.

6. If V is the set of all functions $f : X \to \mathbb{R}$, when for every two functions $f(x)$ and $g(x)$ hold $(f+g)(x) = f(x) + g(x)$, and for every $\alpha \in \mathbb{R}$ there holds $(\alpha f)(x) = \alpha f(x)$, then V is a real vector space when zero vector is a function which is identically zero.

8.2   Sub-spaces

a. Non-empty subset W of V is a **subspace** if, and only if, it is closed under addition and under scalar multiplication.

b. If U and W are subspaces of vector space V, then $W \cap U$ is a subspace.

c. Union of subspaces is not necessarily a subspace.

   Example: $W = \{(a,0,0)\}$ and $U = \{(0,b,0)\}$ are subspaces in $\mathbb{R}^3$. $W \cup U$ is not closed under addition. For example: $(1,0,0) \in W$, $(0,1,0) \in U$, but $(0,1,0) + (1,0,0) = (1,1,0) \notin W \cup U$.

d. Let $\mathbf{v}_1, \mathbf{v}_2, \ldots, \mathbf{v}_n$ be vectors in a vector space over $\mathbb{F}$ and $\alpha_1, \ldots, \alpha_n$ be scalars in $\mathbb{F}$. Vector $\mathbf{v} = \sum_{i=1}^{n} \alpha_i \mathbf{v}_i$ is called a **linear combination** of vectors $\mathbf{v}_1, \mathbf{v}_2, \ldots, \mathbf{v}_n$.

e. Let S be a non empty subset of vector space V. Set $L(S) = \text{span}\{S\}$ of all linear combinations of S is a **span** of S, and $L(\varnothing) = \{\mathbf{0}\}$.

f. If $S \subset V$ is a nonempty subset, then:

   1. $L(S)$ is a subspace in V containing S.

   2. If W is a subspace of V containing S, then $L(S) \subset W$.

8.3   Basis and Dimensions of a Vector Space

a. Vectors $\mathbf{v}_1, \mathbf{v}_2, \ldots, \mathbf{v}_n$ are **linearly dependent** in V if there exist scalars, α_i, $i = 1, 2, \ldots, n$, not all equal to zero, such that $\sum_{i=1}^{n} \alpha_i \mathbf{v}_i = \mathbf{0}$. Otherwise, they are **linearly independent**.

b. Subset S of vector space V is **linearly independent** if it has no finite set of linearly dependent vectors. Otherwise, S is **linearly dependent.**

c. **Definition 1:** A set of vectors S in vector space V is a **basis** if:

   1. It is linearly independent.

   2. V is a span of S.

   Definition 2: A set of vectors S is a basis in V if every vector of V can be **uniquely** presented as a linear combination of the vectors of S.

   Definitions 1 and 2 are equivalent.

d. If vectors $\mathbf{v}_1, \mathbf{v}_2, \ldots, \mathbf{v}_n$ span V and set of vectors $\mathbf{w}_1, \mathbf{w}_2, \ldots, \mathbf{w}_m$ are linearly independent in V, then $m \leq n$.

e. If some basis of vector space V has a finite number of vectors, then any other basis of V has the same number of vectors.

f. Vector space V has **finite dimension** n if its basis consists of n vectors and $n = \dim V$.
   The dimension of null space is $\dim\{\mathbf{0}\} = 0$.

g. Every set of vectors linearly independent in V is either a basis or can be expanded to a basis.

h. Every set of n+1 vectors in an n-dimensional vector space are lineary dependent.

8.4 Sum and Direct Sum of Sub-spaces

a. Sum $U + W$ of subsets U and W of a vector space is the set of **all** vectors $\mathbf{u} + \mathbf{w}$ when $\mathbf{u} \in U$, $\mathbf{w} \in W$.

b. The sum of subspaces is a subspace.

c. If U and W are finite-dimensional subspaces of vector space V, then $U + W$ is finite-dimensional and

$$\dim(U + W) = \dim U + \dim W - \dim(U \cap W)$$

d. Vector space V is a **direct sum** of subspaces U and W if every vector $\mathbf{v} \in V$ can be **uniquely represented** by $\mathbf{v} = \mathbf{u} + \mathbf{w}$, when $\mathbf{u} \in U$, $\mathbf{w} \in W$. It is denoted $V = U \oplus W$.

e. Vector space V is a direct sum of subspaces U and W if and only if $V = U + W$ and $V \cap W = \{\mathbf{0}\}$.

f. Let W_2, W_1 be subspaces of V, S_1 a basis of W_1 and S_2 a basis of W_2. If:

   1. $V = W_1 \oplus W_2$, then $S_1 \cup S_2$ is a basis of V.

2. S_1 and S_2 are linearly independent, and $S_1 \cup S_2$ is a basis of V, then $V = W_1 \oplus W_2$.

g. If $V = W \oplus U$, then $\dim V = \dim U + \dim W$.

h. Vector space U is a **direct sum** of k **subspaces** $W_1, W_2, ..., W_k$ if every vector $\mathbf{v} \in V$ **can be** uniquely represented as $\mathbf{v} = \mathbf{w}_1 + \mathbf{w}_2 + ... + \mathbf{w}_k$ when $\mathbf{w}_i \in W_i, i = 1, 2, ..., k$, it is denoted $V = \oplus_{i=1}^{k} W_i$.

i. Subspace U is called **complement of subspace** W in V if $V = W \oplus U$.

j. Every subspace W of vector space V has a complement in V.

k. A subspace of a vector space has a more than one complement in V.

8.5   Coordinate Vector

a. Let $e = \{\mathbf{e}_1, \mathbf{e}_2, ..., \mathbf{e}_n\}$ be a basis in vector space V over field $\mathbb{F}$. Every vector $\mathbf{v} \in V$ can be represented as $\mathbf{v} = \alpha_1 \mathbf{e}_1 + \alpha_2 \mathbf{e}_2 + ... + \alpha_n \mathbf{e}_n$. This representation is unique. The coefficient vector

$$\mathbf{v}_e = \begin{pmatrix} \alpha_1 \\ \alpha_2 \\ \vdots \\ \alpha_n \end{pmatrix}$$

is called **coordinate vector** of $\mathbf{v}$ in basis e.

b. If $\mathbf{u}, \mathbf{v} \in V$ and e is a basis in V, then $(\mathbf{u} + \mathbf{v})_e = \mathbf{u}_e + \mathbf{v}_e$, $(\alpha \mathbf{u})_e = \alpha \mathbf{u}_e$, $\alpha \in \mathbb{F}$.

8.6 Coordinates of Vector in Various Bases

a. Let $e = \{e_1, e_2, \ldots, e_n\}$ and $f = \{f_1, f_2, \ldots, f_n\}$ be two bases in vector space V. Every vector f_i has a unique representation in basis e.

$$f_i = \sum_{j=1}^{n} a_{ij} e_j \quad , \quad i = 1, 2, \ldots, n$$

The transpose matrix of system

$$P = \begin{pmatrix} a_{11} & a_{21} & \cdots & a_{n1} \\ a_{12} & a_{22} & \cdots & a_{n2} \\ \vdots & \vdots & & \vdots \\ a_{1n} & a_{2n} & \cdots & a_{nn} \end{pmatrix}$$

is called a **transformation matrix** from basis e to basis f.

b. If P is the matrix of transformation from basis e to basis f. then, for every vector $v \in V$ there holds $Pv_f = v_e$, when v_f and v_e are coordinate vectors of v, in bases f and e respectively.

c. The transformation matrix from basis e to basis f is invertible, and $P^{-1}v_e = v_f$.

Chapter 9
Linear Transformations

9.1 Transformations

a. Let X and Y be two nonempty sets. If, for each $x \in X$, there is a unique corresponding element $y \in Y$, it is called **transformation** from X to Y and is denoted $f : X \to Y$ or $f(x) = y$.

x is the **preimage** of y, and y is the **image** of x.

The **image** of transformations is the set of all images of X. It is denoted $\operatorname{Im} f = f(X) = \{f(x) : x \in X\}$.

b. $f : X \to Y$ is **one-to-one correspondence** transformation (denoted $1 \leftrightarrow 1$) if different elements of X have different corresponding images, that is, $x_1 \neq x_2 \Rightarrow f(x_1) \neq f(x_2)$ or $f(x_1) = f(x_2) \Rightarrow x_1 = x_2$.

c. $f : X \to Y$ is **onto** if $\operatorname{Im} f = Y$. In other words, every $y \in Y$ has at least one preimage of X.

d. Let f be a transformation from X to Y, and let S be subset of X. Transformation $f_s : S \to Y$ is a **restruction** of f on S, if its domain of definition is S, in which it is defined exactly like f, that is, for every $x \in S$, $f_S(x) = f(x)$.

9.2   Isomorphism

a. **Definition:** Let V and W be two vector spaces over field $\mathbb{F}$. One-to-one correspondence transformation φ of vector space V over vector space W is called **isomorphism** if for every $\mathbf{u}, \mathbf{v} \in V$ and $k \in \mathbb{F}$, there holds:

$$\varphi(\mathbf{u} + \mathbf{v}) = \varphi(\mathbf{u}) + \varphi(\mathbf{v})$$

$$\varphi(k\mathbf{u}) = k\varphi(\mathbf{u})$$

b. Spaces V and W are **isomorphic** if the exists isomorphism from V to W.

c. If spaces V and W are isomorphic, then, under this isomorphism, a zero of space V is transformed to a zero of space W.

d. If φ is an isomorphism from vector space V to W, then set of vectors $\mathbf{v}_1, \mathbf{v}_2, \ldots, \mathbf{v}_n \in V$ is linearly dependent if, and only if, the set of images $\varphi(\mathbf{v}_1), \varphi(\mathbf{v}_2), \ldots, \varphi(\mathbf{v}_n)$ is linearly dependent on W.

e. If spaces V and W are isomorphic, then a linearly independent set on V transforms under the isomorphism to linearly independent set in W. Therefore, one basis transforms to another basis and $\dim W = \dim V$.

f. If V is an n-dimensional vector space over field $\mathbb{F}$, then space V is isomorphic to $\mathbb{F}^n$.

g. Every two n-dimensional vector spaces over the same field are isomorphic.

9.3   Linear Transformation

a. The transformation f from V into U is called a **linear transformation**, if for every $\mathbf{v}_1, \mathbf{v}_2 \in V$ and $\alpha \in \mathbb{F}$ there holds:

1. $f(\mathbf{v}_1 + \mathbf{v}_2) = f(\mathbf{v}_1) + f(\mathbf{v}_2)$

2. $f(\alpha \mathbf{v}_1) = \alpha f(\mathbf{v}_1)$

b. $f(\mathbf{0}_v) = \mathbf{0}_u$. That is, in a linear transformation, the image of zero $\mathbf{0}_v$ of V is (zero) $\mathbf{0}_u$ of U.

c. An **identity transformation** I_v, which transforms every vector in V to itself, is a linear transformation from V on V.

d. A **zero transformation** $0 : V \rightarrow U$, which transforms every vector of V to zero vector of U, is a linear transformation from V to U.

e. If $e = \{\mathbf{v}_1, \mathbf{v}_2, \ldots, \mathbf{v}_n\}$ is a basis of V and $\mathbf{u}_1, \mathbf{u}_2, \ldots, \mathbf{u}_n$ are vectors (not necessarily linearly independent) in space U, then there exists a **unique** linear transformation T from V to U, such that $T(\mathbf{v}_i) = \mathbf{u}_i$, $i = 1, 2, \ldots, n$.

9.4   Image and Kernel of Linear Transformation

a. If $\{\mathbf{v}_1, \mathbf{v}_2, \ldots, \mathbf{v}_n\}$ is a basis in vector space V and T is a linear transformation from V to U, then the image of transformation T is:

$$\operatorname{Im} T = \operatorname{span}\left(T(\mathbf{v}_1), T(\mathbf{v}_2), \ldots, T(\mathbf{v}_n)\right)$$

b. The **Kernel**, $\operatorname{Ker} T$, of linear transformation $T : V \rightarrow U$ is the set of vectors in V the images of which are the zero vector of U,

$$\operatorname{Ker} T = \{\mathbf{v} \in V : T(\mathbf{v}) = \mathbf{0}_u\}$$

c. If $T : V \rightarrow U$ is a linear transformation, then:

1. $\operatorname{Im} T$ is a subspace of U.

2. $\operatorname{Ker} T$ is a subspace of V.

d. If $T : V \rightarrow U$ is a linear transformation and V is an n-dimensional vector space, then

$$\dim V = \dim \operatorname{Ker} T + \dim \operatorname{Im} T$$

e. $\dim \operatorname{Im} T$ is the **transformation rank** of T ($\operatorname{rank} T$).

9.5   Linear Operator

a. A **linear operator** on V is a linear transformation from V to itself.

b. Linear operator $T : V \to V$ is **non-singular** if $\operatorname{Ker} T = \{0\}$. Otherwise, T is a **singular operator**.

c. Linear operator $T : V \to V$ is one-to-one correspondence if, and only if, T is non-singular.

d. If T and S are linear operators on V, then operators $T + S$ and $T \cdot S$, defined as $(T + S) v = T v + S v$ and $(T \cdot S) v = T(S v)$, respectively, are linear operators on V.

e. Linear operator $T : V \to V$ is called **invertible** if there exists operator T^{-1} such that $T T^{-1} = T^{-1} T = I$.

f. If T and S are invertible operators, then operator $S \cdot T$ is invertible, and $(TS)^{-1} = S^{-1} \cdot T^{-1}$.

g. If T is a linear operator in an n-dimensional vector space, the following propositions are equivalent:

   1. T is one-to-one correspondent.

   2. T is non-singular.

   3. T is onto.

   4. T is invertible.

9.6   Matrix Representation of Linear Operator

a. Let $T : V \to V$ and let $v = \{v_1, v_2, \ldots, v_n\}$ be a basis in V. We represent

$$T v_j = \sum_{i=1}^{n} a_{ij} v_i , \quad j = 1, 2, \ldots, n$$

The n-th order square matrix

$$T_v = \begin{pmatrix} a_{11} & a_{12} & \cdots & a_{1n} \\ a_{21} & a_{22} & \cdots & a_{2n} \\ \vdots & \vdots & & \vdots \\ a_{n1} & a_{n2} & \cdots & a_{nn} \end{pmatrix}$$

is the **representative matrix** of operator T with respect to basis V, or that T_v is a **matrix representation** of T on V.

b. If $T:V \to V$ is a linear operator and $v = \{v_i\}_{i=1}^{n}$ is a basis of V, then, for every vector $u \in V$, there holds $T_v u_v = (Tu)_v$.

c. Let e and f be bases in vector space V. If P is a matrix of transformation from e to f, then, for every linear operator $T:V \to V$, there holds

$T_f = P^{-1} T_e P$

when T_f, T_e are matrix representations of T, with respect to bases e and f.

Chapter 10
Matrix Similarity

a. Square matrices A and B in $\mathbb{F}^{n \times n}$ are **similar** if there exists invertible matrix P in $\mathbb{F}^{n \times n}$, such that $B = P^{-1}AP$. It is denoted $A \sim B$.

b. Properties of similarity: If $A \sim B$

1. $B \sim A$

2. $A^n \sim B^n$.

3. $\det A = \det B$

4. $\text{tr}A = \text{tr}B$

5. $\text{rank}A = \text{rank}B$

c. Matrices A and B represent the same linear operator in different bases if, and only if, they are similar.

d. $A \in \mathbb{F}^{n \times n}$ is a **diagonalizable matrix** if it is similar to a diagonal matrix, that is, if there exists invertible matrix P and diagonal matrix $D = \text{diag}\{d_1, \ldots, d_n\}$ such that $P^{-1}AP = D$.

e. If A is similar to $\text{diag}(d_1, d_2, \ldots, d_n)$, then A is similar to $\text{diag}(d_{i_1}, d_{i_2}, \ldots, d_{i_n})$ for every permutation $(i_1, i_2, \ldots, i_n)$.

Chapter 11
Eigenvalues and Eigenvectors

11.1 Definitions; Diagonalization of Matrices

a. Non-zero vector $\mathbf{x} \in \mathbb{F}^n$ is an **eigenvector** of matrix $A \in \mathbb{F}^{n \times n}$ if it is proportionate (parallel) to vector $A\mathbf{x}$, that is

$$A\mathbf{x} = \lambda \mathbf{x}$$

Scalar λ is the **eigenvalue** corresponding to eigenvector $\mathbf{x}$.

b. Matrix $A \in \mathbb{F}^{n \times n}$ is diagonizable if and only if A has n linearly independent eigenvectors. Furthermore, if P is a matrix in which the columns are eigenvectors of A, then, the diagonal of diagonal matrix $P^{-1}AP = D$ will have the corresponding eigenvalues, when the order of diagonal elements in D corresponds to the order of rows in P.

11.2 Characteristic Polynomials

a. Polynomial $\Delta_A(t) = \det(tI - A)$ is   a   **characteristic polynomial** of A and matrix $tI - A$ is a **characteristic matrix** of A.

b. The roots of characteristic polynomial are the eigenvalues of matrix A.

c. The **characteristic polynomial of a finite dimension linear operator** is the characteristic polynomial of its representative matrix.

d. Eigenvectors of matrix A corresponding to different eigenvalues are linearly independent.

e. If matrix $A \in \mathbb{F}^{n \times n}$ has n different eigenvalues, then A is diagonizable.

11.3 Eigenvalues of Operators

Let $T : V \rightarrow V$ be a linear operator.

a. Vector **v** is an **eigenvector** of operator T if $\mathbf{v} \neq \mathbf{0}$ and if there exists scalar λ, such that $T\mathbf{v} = \lambda\mathbf{v}$. Scalar λ is called **eigenvalue** of T corresponding to the eigenvector **v** (**v** is an eigenvector corresponding to λ).

b. The kernel of $T - \lambda I$ is called an **eigenspace** of T corresponding to eigenvalue λ and is denoted V_λ.

c. T is diagonizable if, and only if, V has a basis **v** constructed of eigenvectors of T. In such a case, $T_v = \text{diag}\{d_1,\ldots,d_n\}$, when d_i is the eigenvalue of T corresponding to $\mathbf{v}_i$.

11.4 Invariant Subspaces

a. Let T is linear operator over vector space V, and let W be a subspace of V and T_W be a restriction of T over W, then subspace W is T-**invariant** of V if $T_W \subset W$.

b. $\{0\}$ and V are invariant subspaces of V under every linear operator in V.

c. If λ is an eigenvalue of T then the corresponding eigenspace V_λ is T-**invariant**.

d. If W is a T-invariant subspace of V and $v = \{\mathbf{v}_1, ..., \mathbf{v}_r, ..., \mathbf{v}_n\}$ is a basis of V, and $\{\mathbf{v}_1, ..., \mathbf{v}_r\}$ is a basis of W, then T_v is a block triangular matrix

$$T_v = \begin{pmatrix} B & C \\ 0 & D \end{pmatrix}$$ where $B \in \mathbb{C}^{r \times r}$, $D \in \mathbb{C}^{(n-r) \times (n-r)}$, and B is the

matrix representing the restriction of T over W.

e. If T is a diagonizable operator over a finite-dimension vector space V, $\lambda_1, ..., \lambda_k$ are the eigenvalues of T, and $V_1, ..., V_k$ are the corresponding eigenspaces, then V is the direct sum of the eigenspaces $V = V_1 \oplus ... \oplus V_k$.

11.5 Geometric and Algebraic Multiplicity of Eigenvalues

a. The **geometric multiplicity** of eigenvalue λ of operator T is a dimension of the eigenspace of T corresponding to λ.

b. The **algebraic multiplicity** of eigenvalue λ of operator T is the multiplicity of λ as a root of the characteristic polynomial $\Delta_T(\lambda)$.

c. Matrix $\Delta_T(\lambda)$ has geometric multiplicity of λ equal to $n - \text{rank}(A - \lambda I)$.

d. The geometric multiplicity of eigenvalue λ does not exceed the algebraic multiplicity of λ.

e. $A \in \mathbb{F}^{n \times n}$ is diagonizable if and only if every eigenvalue λ of A has a geometric multiplicity equal to its algebraic multiplicity.

11.6 Cayly-Hamilton Theorem

a. Let $f(t) = a_k t^k + a_{k-1} t^{k-1} + \ldots + a_1 t + a_0$ be a polynomial of degree k with coefficients of field $\mathbb{F}$ and $A \in \mathbb{F}^{n \times n}$. Then, the polynomial in A

$$f(A) = a_k A^k + a_{k-1} A^{k-1} + \ldots + a_1 A + a_0 I$$

is a matrix of the following properties:

1. If $Ax = \lambda x$ then, for every polynomial $f(t)$ there holds $f(A)x = f(\lambda)x$.

2. If there exists invertible matrix P such that $P^{-1}AP = D$ is a diagonal matrix, then, for that P holds $f(A) = Pf(D)P^{-1}$.

b. **Cayley-Hamilton Theorem**: If A is a square matrix and $\Delta_A(x)$ its characteristic polynomial, then $\Delta_A(A) = 0$.

c. If A is invertible, then A^{-1} is the polynomial on A.

11.7 Minimal Polynomial

a. The polynomial equation of the lowest degree which A satisfies is called minimal polynomial and is denoted as $m_A(t)$.

b. For every polynomial $f(t)$ which satisfies $f(A) = 0$, minimal polynomial $m_A(t)$ is a divider of $f(t)$ ($m \mid f$).

c. Every square matrix A has a unique minimal polynomial.

d. Every root of the minimal polynomial is an eigenvalue.

e. $m_A(t)$ and $\Delta_A(t)$ have the same roots.

f. Similar matrices have the same minimal polynomial.

g. Let A be **block diagonal matrix** $A = \operatorname{diag}(A_1, A_2, \ldots, A_k)$, where the main diagonal of A has square matrices $A_1, A_2, \ldots, A_k$. Then $m_A(t)$ is the least common product of the corresponding minimal polynomials $m_{A_1}(t), m_{A_2}(t), \ldots, m_{A_k}(t)$.

11.8 Spectral Factorization

a. Transformation $E: V \to U$ is called the projection on U parallel to W if $V = U \oplus W$ (see 8.4).

b. Operator E holding $E^2 = E$ is called **projection operator**.

c. **Spectral Factorization Theorem:** Let T be a diagonizable operator over n-dimensional vector space, V. Let $\lambda_1, \ldots, \lambda_k$ be various eigenvalues of T $(k \le n)$, and let $V_1, \ldots, V_k$ be he eigenspaces of $\lambda_1, \ldots, \lambda_k$, respectively, then:

1. $V = V_1 \oplus V_2 \oplus \ldots \oplus V_k$.

2. If E_i is the projection of V_i parallel to the direct sum of the rest of eigenspaces, then $E_1 + E_2 + \ldots + E_k = I$. The representation $T = \lambda_1 E_1 + \lambda_2 E_2 + \ldots + \lambda_k E_k$ is the **spectral factorization** of T.

3. For every polynomial f ,

$$f(T) = f(\lambda_1)E_1 + f(\lambda_2)E_2 + \ldots + f(\lambda_k)E_k$$

4. $E_j = e_j(T)$,   when $e_i(x) = \displaystyle\prod_{i \neq j} \frac{x - \lambda_i}{\lambda_j - \lambda_i}$ (polynomials $e_j(x)$ are called **Lagrange Polynomials**).

5. A factorization where the λ_i are different, $E_i^2 = E_i$, $E_i E_j = 0$ for every $i \neq j$, is unique.

11.9 Jordan Form

a. **Jordan block**, $J(a;k)$, is a $k \times k$-order matrix where the diagonal elements equal to a the elements above the diagonal parallel to it equal 1, and the rest of the coefficients equal 0.

b. The characteristic polynomial and minimal polynomial of $J(a,k)$ are $(t-a)^k$. The algebraic multiplicity of a on $J(a;k)$ is k, and its geometric multiplicity is 1.

c. A matrix is of **Jordan form** if it is a block diagonal matrix where the diagonal cells are Jordan blocks.

d. If $A \in \mathbb{F}^{n \times n}$ can be reduced to triangular form, then A is similar to a matrix of a Jordan form. This matrix is unique up to the order of blocks.

e. If $J = \text{diag}\{J_1, \ldots, J_r\}$ when $J_i = J(a_i, k_i)$ are Jordan blocks, then, for every matrix A similar to J :

1. $\Delta_A(t) = (t-a_1)^{k_1} \cdot (t-a_2)^{k_2} \cdot \ldots \cdot (t-a_r)^{k_r} = \displaystyle\prod_{i=1}^{r} (t-a_i)^{k_i}$

2. For every eigenvalue a , the power of $(t-a)$ in $m_A(t)$ is the highest Jordan block order corresponding to a .

3. The geometric multiplicity of eigenvalue a is the number of blocks corresponding to it.

Chapter 12
Quadratic Form

a. The expression $f(x) = x^t A x$, when A is a symmetric matrix in $\mathbb{R}^{n \times n}$ and $x \in \mathbb{R}^n$ is called **quadratic form**.

b. Example in $\mathbb{R}^3$:

$$f(x) = (x_1 x_2 x_3) \begin{pmatrix} a_{11} & a_{12} & a_{13} \\ a_{12} & a_{22} & a_{23} \\ a_{13} & a_{23} & a_{33} \end{pmatrix} \begin{pmatrix} x_1 \\ x_2 \\ x_3 \end{pmatrix} = a_{11}x_1^2 + a_{22}x_2^2 + a_{33}x_3^2 +$$

$$+2a_{12}x_1x_2 + 2a_{13}x_1x_3 + 2a_{23}x_2x_3$$

A quadratic form is positive (non-negative) if for every

$$x \in \mathbb{R}^n, \ x^t A x > 0 \ \ (x^t A x \geq 0)$$

c. A quadratic form is negative (non-positive) if for every $x \in \mathbb{R}^n$, $x^t A x < 0$ $(x^t A x \leq 0)$.

d. The following propositions are equivalent

1. Quadratic form $f(x)$ is positive.

2. The eigenvalues of A are positive.

3. All numbers

$$a_{11}, \ \begin{vmatrix} a_{11} & a_{12} \\ a_{12} & a_{22} \end{vmatrix}, \ \begin{vmatrix} a_{11} & a_{12} & a_{13} \\ a_{12} & a_{22} & a_{23} \\ a_{13} & a_{23} & a_{33} \end{vmatrix}, \dots, \det A \qquad (*)$$

are positive.

e. Quadratic form $f(x)$ is negative if and only if the signs of the n numbers in (*) change intermittently, the first a_{11} being negative.

Chapter 13
Inner Product Spaces

13.1 Inner Product

a. V is an **inner product space** over field $\mathbb{F}$ if every ordered pair of vectors $\mathbf{a}, \mathbf{b} \in V$ has a corresponding number of $\mathbb{F}$ called **the inner product** of vectors $\mathbf{a}$ and $\mathbf{b}$ and denoted $\langle \mathbf{a}, \mathbf{b} \rangle$ such that:

1. $\langle \mathbf{a}, \mathbf{b} \rangle = \overline{\langle \mathbf{b}, \mathbf{a} \rangle}$

2. $\langle \alpha \mathbf{a}, \mathbf{b} \rangle = \alpha \langle \mathbf{a}, \mathbf{b} \rangle$, $\alpha \in \mathbb{F}$

3. $\langle \mathbf{a} + \mathbf{b}, \mathbf{c} \rangle = \langle \mathbf{a}, \mathbf{c} \rangle + \langle \mathbf{b}, \mathbf{c} \rangle$

4. $\langle \mathbf{a}, \mathbf{a} \rangle > 0$ for every $\mathbf{a} \neq \mathbf{0}$.

b. If $\mathbb{F}$ is a field of complex numbers $\mathbb{C}$, then V is a **unitary space**.

c. If $\mathbb{F}$ is a field of real numbers $\mathbb{R}$, then V is a **Euclidean space**.

d. In a Euclidean space, $\langle \mathbf{a}, \mathbf{b} \rangle = \langle \mathbf{b}, \mathbf{a} \rangle$.

e. Every finite-dimensional vector space V over fiend $\mathbb{F}$ is unitary.

f. **Examples**:

1. Vector space $\mathbb{C}^n$ with inner product $\langle \mathbf{a}, \mathbf{b} \rangle = \sum\limits_{i=1}^{n} a_i \overline{b}_i$.

2. A space of continuous real functions on interval $[a, b]$ with the inner product $\langle a(t), b(t) \rangle = \int\limits_{a}^{b} a(t) b(t) \, dt$ is a Euclidean space.

3. Complex matrices space $\mathbb{C}^{n \times n}$ is a unitary space with the inner product

$$\langle A, B \rangle = \mathrm{tr}(\overline{B^t A}) = \sum_{i=1}^{m} \sum_{j=1}^{n} a_{ij} \overline{b}_{ij} \ , \quad A, B \in \mathbb{C}^{n \times n}$$

4. ℓ^2 is the space of all infinite sequences $(a_1, a_2, \ldots)$ such that the series $\sum\limits_{i=1}^{\infty} a_i^2$ is convergent. The inner product of two vectors of ℓ^2 is

$$\langle (a_1, a_2, \ldots), (b_1, b_2, \ldots) \rangle = \sum_{i=1}^{\infty} a_i b_i$$

13.2 Cauchy-Schwartz Inequality

a. Non-negative number $\| \mathbf{a} \| = \sqrt{\langle \mathbf{a}, \mathbf{a} \rangle}$ is called a **length** or a **norm** of vector $\mathbf{a}$.

b. For every vectors $\mathbf{a}, \mathbf{b}$ of a unitary space, there holds:

$$| \langle \mathbf{a}, \mathbf{b} \rangle | \leq \| \mathbf{a} \| \cdot \| \mathbf{b} \|$$

This equality holds if and only if $\mathbf{a}$ and $\mathbf{b}$ are linearly dependent.

c. In a continuous functions space (see example f.2), Cauchy-Schwartz inequality is in the form of

$$\left| \int\limits_a^b f(t)g(t)\,dt \right|^2 \le \int\limits_a^b [f(t)]^2\,dt \cdot \int\limits_a^b [g(t)]^2\,dt$$

d. **Triangle inequality:** The length of sum of two vectors is no greater than the sum of their lengths. That is, $\|\mathbf{a}+\mathbf{b}\| \le \|\mathbf{a}\| + \|\mathbf{b}\|$, and this equality holds if, and only if, $\mathbf{a}=\alpha\mathbf{b}, \alpha \ge 0$ (that is, $\mathbf{a}$ and $\mathbf{b}$ are parallel and in the same direction).

e. Angle φ between non-zero vectors $\mathbf{a},\mathbf{b}$ of vector field V is defined by $\cos\varphi = \dfrac{\langle \mathbf{a},\mathbf{b}\rangle}{\|\mathbf{a}\| \cdot \|\mathbf{b}\|}$, $0 \le \varphi \le \pi$.

f. Angle φ exists and is unique.

13.3 Orthogonality

a. Vectors $\mathbf{a}$ and $\mathbf{b}$ of unitary space V are **orthogonal (perpendicular)** if their inner product is equal to zero, that is, $\langle \mathbf{a},\mathbf{b}\rangle = 0$.

b. Vector $\mathbf{0}$ is the only vector that is orthogonal to any other vector of V .

c. Set of vectors S of unitary space V is **orthogonal** if any two different vectors of S are orthogonal.

d. An orthogonal set S of non-zero vectors is linearly independent. If, in addition, it spans the space, then it is an **orthogonal basis** in span S .

e. **Gram-Schmidt theorem:** every n -dimensional unitary space V has an orthogonal basis. Moreover, for every linearly independent set spanning a subspace W , there is an orthogonal set spanning W .

f. **Gram-Schmidt orthogonalization**: If $S=\{\mathbf{a}_1,\mathbf{a}_2,...,\mathbf{a}_n\}$ is linearly independent set, we construct a set of n orthogonal non-zero vectors $\mathbf{b}_1, \mathbf{b}_2, \ldots, \mathbf{b}_n$ such that for every $1 \le p \le n$, $\operatorname{span}\{\mathbf{b}_1, \ldots, \mathbf{b}_p\} = \operatorname{span}\{\mathbf{a}_1, \ldots, \mathbf{a}_p\}$ the following way:

1. Define $\mathbf{b}_1 = \mathbf{a}_1$.

2. Choose k_1 such that $\mathbf{b}_2 = k_1 \mathbf{b}_1 + \mathbf{a}_2$ is orthogonal to $\mathbf{b}_1$.

The result is $k_1 = -\dfrac{\overline{\langle \mathbf{b}_1, \mathbf{a}_2 \rangle}}{\|\mathbf{b}_1\|^2} = -\dfrac{\langle \mathbf{a}_2, \mathbf{b}_1 \rangle}{\|\mathbf{b}_i\|^2}$.

Therefore, $\mathbf{b}_2 = \mathbf{a}_2 - \dfrac{<\mathbf{a}_2, \mathbf{b}_1>}{\|\mathbf{b}_i\|^2} \mathbf{b}_1$.

The same way we get

$$\mathbf{b}_p = \mathbf{a}_p - \sum_{i=1}^{p-1} \frac{\langle \mathbf{a}_p, \mathbf{b}_i \rangle}{\|\mathbf{b}_i\|^2} \mathbf{b}_i \ , \ p = 2, \ldots, n$$

13.4 Orthonormal Basis

a. An orthogonal vector set is **orthonormal** if every vector of the set is normalized (that is, it has a unit length).

b. Every unitary space has an orthonormal basis.

c. In unitary space V, the components of a vector $\mathbf{v}$ in orthonormal basis $\mathbf{e}_1, \mathbf{e}_2, \ldots, \mathbf{e}_n$ are

$$\mathbf{v}_e = \left(\langle \mathbf{v}, \mathbf{e}_1 \rangle, \langle \mathbf{v}, \mathbf{e}_2 \rangle, \ldots, \langle \mathbf{v}, \mathbf{e}_n \rangle \right)$$

$$\mathbf{v} = \sum_{i=1}^{n} \langle \mathbf{v}, \mathbf{e}_i \rangle \mathbf{e}_i$$

d. If $e = \{\mathbf{e}_1, \mathbf{e}_2, \ldots, \mathbf{e}_n\}$ is a basis in unitary space V and $\mathbf{v}_e = (\alpha_1, \alpha_2, \ldots, \alpha_n)$, $\mathbf{w}_e = (\beta_1, \beta_2, \ldots, \beta_n)$ are two vectors of V, then basis e is orthonormal if and only if the inner product of **every two** vectors $\mathbf{v}$ and $\mathbf{w}$ is $\langle \mathbf{v}, \mathbf{w} \rangle = \alpha_1 \bar{\beta}_1 + \alpha_2 \bar{\beta}_2 + \ldots + a_n \bar{\beta}_n$.

13.5   Fourier Coefficients

a. Let $e = \{\mathbf{e}_1, \mathbf{e}_2, \ldots\}$ orthonormal basis in unitary space V and $\mathbf{v} \in V$. Scalars $\alpha_i = \langle \mathbf{v}, \mathbf{e}_i \rangle, i = 1, 2, \ldots$ are called **Fourier coefficients** of $\mathbf{v}$ in respect to e.

b. Let V be an inner product space and $\mathbf{u}, \mathbf{v} \in V$. The distance between $\mathbf{u}$ and $\mathbf{v}$ is non-negative number $\| \mathbf{u} - \mathbf{v} \|$.

c. If $\{\mathbf{e}_1, \mathbf{e}_2, \ldots, \mathbf{e}_n\}$ is an orthonormal system in vector space V $W = \mathrm{span}\{\mathbf{e}_1, \mathbf{e}_2, \ldots, \mathbf{e}_n\}$ and $\mathbf{u} \in V$,   then   vector $\tilde{\mathbf{u}} = \sum_{i=1}^{n} \langle \mathbf{u}, \mathbf{e}_i \rangle \mathbf{e}_i$ is the closest vector to $\mathbf{u}$ of W. Moreover, $\tilde{\mathbf{u}}$ is the unique vector of W at a minimum distance from $\mathbf{u}$.

d. **Bessel's Inequality:** If $\{\mathbf{e}_1, \mathbf{e}_2, \ldots, \mathbf{e}_n\}$ is an orthonormal system on V, then, for every vector $\mathbf{v} \in V$ there holds:

$$\sum_{i=1}^{n} \left| \langle \mathbf{v}, \mathbf{e}_i \rangle \right|^2 \le \| \mathbf{v} \|^2$$

Equality holds if, and only if, $V = \mathrm{span}\{\mathbf{e}_1, \mathbf{e}_2, \ldots, \mathbf{e}_n\}$.

This is called **Parseval's equality**.

13.6   Infinite Orthonormal System

Let V be an inner product space and $\{\mathbf{e}_1, \mathbf{e}_2, \ldots, \}$ an infinite orthonormal system.

a. **Bessel's Inequality:** for every $\mathbf{v} \in V$, series $\sum_{n=1}^{\infty} \left| \langle \mathbf{v}, \mathbf{e}_n \rangle \right|^2$ converges and there holds

$$\sum_{n=1}^{\infty} |<\mathbf{v},\mathbf{e}_n>|^2 \le \|\mathbf{v}\|^2$$

b. Let $\{\mathbf{w}_m\}_{m=1}^{\infty}$ be an infinite series of vectors in a normed space V. This sequence is **convergent in norm** to vector $\mathbf{w} \in V$ if $\lim_{m\to\infty} \|\mathbf{w}-\mathbf{w}_m\|=0$. Which means, for each $\varepsilon>0$ there exists integer $m(\varepsilon)$ such that for each $m \ge m(\varepsilon)$, $\|\mathbf{w}-\mathbf{w}_m\|<\varepsilon$ holds.

c. **Definition**: Let $\{\mathbf{u}_1,\mathbf{u}_2,...\}$ be an infinite sequence of vectors in a normed space and let $\{a_n\}_{n=1}^{\infty}$ be a scalar sequence. Series $\sum_{n=1}^{\infty} a_n\mathbf{u}_n$ is said to be **convergent in norm** to vector $\mathbf{w} \in V$, is denoted $\mathbf{w} = \sum_{n=1}^{\infty} a_n\mathbf{u}_n$, if the **partial sums** sequence $\mathbf{w}_m = \sum_{n=1}^{m} a_n\mathbf{u}_n$ converges in norm to $\mathbf{w}$. In other words, series $\sum_{n=1}^{\infty} a_n\mathbf{u}_n$ converges in norm to vector $\mathbf{w}$ if $\lim_{m\to\infty} \left\| \mathbf{w}-\sum_{n=1}^{m} a_n\mathbf{u}_n \right\|=0$.

d. The proposition "vector $\mathbf{w}$ is spanned by infinite sequence $\{\mathbf{u}_1,\mathbf{u}_2,...\}$" means there is a matching sequence of scalars $\{a_n\}_{n=1}^{\infty}$ such that as m increases, the combination $a_1\mathbf{u}_1+...+a_m\mathbf{u}_m$ becomes an increasingly better approximation to vector $\mathbf{w}$. The approximation between vectors in a normed space is measured by their distance, and consequently, the exact meaning of the last proposition is that for every $\varepsilon>0$, as small as we wish, there exists an $m(\varepsilon)$ such that for all $m \ge m(\varepsilon)$.

$$\|\mathbf{w}-(a_1\mathbf{u}_1+a_2\mathbf{u}_2+...+a_m\mathbf{u}_{m)}\|<\varepsilon$$

e. **Definition:** Let $\{e_1, e_2, \ldots\}$ be an infinite orthonormal system in an inner product space V. It is **close** in V if for every $u \in V$ there holds

$$\lim_{m \to \infty} \left\| u - \sum_{n=1}^{m} \langle u, e_n \rangle e_n \right\| = 0$$

f. Orthonormal system $\{e_1, e_2, \ldots\}$ is closed in inner product space V if, and only if, for every vector $u \in V$ there holds

$$\sum_{n=1}^{\infty} \left| \langle u, e_n \rangle \right|^2 = \| u \|^2$$

It means that the closeness is equivalent to Parseval's equality for every vector $u \in V$.

g. Orthonormal system $\{e_1, e_2, \ldots\}$ is **complete** in V if the only unique vector holding $\langle u, e_n \rangle = 0$, $\forall n \in \mathbb{N}$ is zero vector $u = 0$.

h. **A Generalization of Parseval's Equality**: If $\{e_1, e_2, \ldots\}$ is a complete orthonormal system in inner product space V, then, for every pair of vectors $u, v \in V$ there holds

$$\langle u, v \rangle = \sum_{n=1}^{\infty} a_n \overline{b}_n \text{ , when } a_n = \langle u, e_n \rangle \text{ and } b_n = \langle v, e_n \rangle$$

XI. Ordinary Differential Equations, or ODE

Chapter 1
Classification of First-Order Ordinary Differential Equations

1.1   Introduction

a. An ordinary differential equation, or ODE, expresses the relation between a one-variable function and its derivatives.
The general form of n-th **order** ODE for function $y(x)$ is

$$F(x, y, y', \ldots, y^{(n)}) = 0$$

when F is a given function of $n + 2$ variables.

b. Function $y = y(x)$ is a **solution** of n-th-order ODE on interval (a, b) if it is n-times differentiable and satisfies the equation

$$F\left(x, y(x), y'(x), \ldots, y^{(n)}(x)\right) \equiv 0$$

c. A first-order ODE is the equation $F(x, y, y') = 0$ or, in its explicit form, is $y' = f(x, y)$.

d. **Example**: The simplest ODE is $y' = f(x)$, and its solutions are $y(x) = \int f(x)\,dx + c$ when c is a constant.

e. **A general solution** of a first-order ODE includes all solutions and is dependent on a constant c. Every specific choice of c, results in a **particular solution.**
The particular solution of $y' = f(x)$ that satisfies $y_0 = y(x_0)$ is $y = y_0 + \int\limits_{x_0}^{x} f(x)\,dx$.

f. The problem consisting of equation $y' = f(x, y)$ and initial conditions $y(x_0) = y_0$ is called **a Cauchy problem**.

1.2   Separable Equations

a. Equations $y' = f(x) \cdot g(y)$, $f(x)g(y)\,dx = \varphi(x)h(y)\,dy$ have separate variables. Such an equation is called **separable equation.**

b. A general solution of $y' = f(x)g(y)$ is

$$\int \frac{dy}{g(y)} = \int f(x)\,dx + c$$

c. A general solution of the equation
$f(x)g(y)\,dx = \varphi(x)h(y)\,dy$ is

$$\int \frac{f(x)}{\varphi(x)}\,dx = \int \frac{h(y)}{g(y)}\,dy + c$$

Division of both sides by $\varphi(x)g(y)$ usually loses **particular solutions** holding $\varphi(x)g(y) = 0$.

d. The equation $y' = f(ax + by)$ is not separable, but turns into separable equation $\dfrac{dz}{dx} = a + bf(z)$ after substituting $z = ax + by$ and its solutions are

$$x + c = \int \frac{dz}{a + bf(z)} \cdot$$

1.3   Homogeneous Equations

a. $g(x, y)$ is a k-**th-order homogeneous function** if for every real t $g(tx, ty) = t^k g(x, y)$.

   Example: Function $f(x, y) = \dfrac{x^2 + xy}{2x^2 + y^2}$ where all the powers of the numerator and of the denominator are equal is a 0-order homogeneous function.

b. ODE $y' = f(x, y)$ is **homogeneous** if $f(x, y)$ is 0-order homogeneous function.

c. Substituting $y = zx$ transforms the homogeneous equation into the separable equation $z'x + z = g(z)$.

d. An equation in the form of $y' = f\left(\dfrac{a_1 x + b_1 y + c_1}{a_2 x + b_2 y + c_2}\right)$ turns into homogeneous equation if the coordinates are shifted to point (x_0, y_0), the intersection of lines

$$a_1 x + b_1 y + c_1 = 0 \quad , \quad a_2 x + b_2 y + c_2 = 0 .$$

After substituting $X = x - x_0$, $Y = y - y_0$, we obtain homogeneous equation

$$\frac{dY}{dX} = f\left(\frac{a_1 X + b_1 Y}{a_2 X + b_2 Y}\right)$$

1.4 Exact Equations

a. The equation $M(x,y)dx + N(x,y)dy = 0$, $(x,y) \in D$ when

$\dfrac{\partial M}{\partial y} = \dfrac{\partial N}{\partial x}$ in domain D is called **exact equation**. In this

case there exist continuous function $u(x,y)$ with continuous partial derivatives, such that $du(x,y) = M(x,y)dx + N(x,y)dy$ and the general solution of the exact equation is $u(x,y) = c$.

b. If $(x_0, y_0) \in D$, then $u(x,y) = \displaystyle\int_{x_0}^{x} M(x,y)dx + \int_{y_0}^{y} N(x_0,y)dy$

or $u(x,y) = \displaystyle\int_{y_0}^{y} N(x,y)dy + \int_{x_0}^{x} M(x,y_0)dx$

1.5 Integrating Factor

a. If the equation $M(x,y)dx + N(x,y)dy = 0$ is not exact, but, if multiplied it by function $\mu(x,y)$, the result is the equation

$$\mu(x,y)M(x,y)dx + \mu(x,y)N(x,y)dy = 0$$

which is an exact equation, that is

$$\frac{\partial}{\partial y}[\mu(x,y)M(x,y)] = \frac{\partial}{\partial x}[\mu(x,y)N(x,y)]$$

Then, function $\mu(x,y)$ is called an **integrating factor**.

b. If the expression $\dfrac{M'_y - N'_x}{N}$ is a function of x only, then

the integrating factor is a function dependent of x only, and is the solution of the equation

$$\frac{\mu'(x)}{\mu(x)} = \frac{M'_y - N'_x}{N}$$

c. If the expression $\dfrac{N'_x - M'_y}{M}$ is a function of y only, then the integrating factor of the equation is dependent of y only, and is the solution of the equation

$$\frac{\mu'(y)}{\mu(y)} = \frac{N'_x - M'_y}{M}$$

d. If the expression $\dfrac{N'_x - M'_y}{xM - yN}$ is a function of $x \cdot y$ only, then the integrating factor is a function dependent of $x \cdot y$, and is the solution of the equation $\dfrac{\mu'(z)}{\mu(z)} = \dfrac{N'_x - M'_y}{xM - yN}$ when $z = x \cdot y$.

1.6　First-Order Linear Equations

a. The **ODE**

$$y' + P(x)y = f(x) \quad (*)$$

when $P(x)$ and $f(x)$ are continuous functions on (α, β) is called **a first-order linear equations.**

b. If $f(x) \equiv 0$, the equation is a **homogeneous linear equation.**

c. General solution of homogeneous linear equation $y' + P(x)y = 0$ is $y_0 = Ce^{-\int P(x)dx}$.

d. $y = C_1 e^{-\int P(x)dx} + e^{-\int P(x)dx} \int f(x)e^{\int P(x)dx}dx$ is a general solution of (*).

e. **Existence and Uniqueness Theorem**: If functions $P(x)$ and $f(x)$ are continuous in open interval (α, β), then there exists a unique function $y = y(x)$ holding (*) and $y(x_0) = y_0$ for every given y_0.

f. **Bernoulli's   Equation**: $y' + P(x)y = f(x)y^n$, $n \neq 1$.
Substituting $z = y^{1-n}$, we get a linear equation

$$\frac{dz}{dx} + (1-n)P(x)z = (1-n)f(x)$$

g. **Riccati Equation**: $y' + P(x)y = g(x)y^2 + f(x)$.
If one particular solution of equation $y_1(x)$ is known,
then by substituting $y(x) = y_1(x) + z(x)$ we get a Bernouli
equation for $z(x)$.

1.7   Existence and Uniqueness Theorem for First-Order ODE

Given an ODE with initial conditions (Cauchy's problem):

$$y' = f(x,y), \ y(x_0) = y_0 \quad (*)$$

a. The integral form of (*):

$$y(x) = y(x_0) + \int_{x_0}^{x} f(t, y(t)) dt$$

b. **Existence and Uniqueness Theorem**: if function $f(x,y)$
is continuous on rectangle

$$R: \{x_0 - a \leq x \leq x_0 + a, y_0 - b \leq y \leq y_0 + b\}$$

and satisfy Lipschitz continuity criteria in R

$$|f(x,y_1) - f(x,y_2)| \leq N|y_1 - y_2|$$

when N is constant and $|f(x,y)| \leq M$ on R, then (*) has
a unique solution $y = y(x)$ on $|x - x_0| < \alpha$ when
$\alpha = \min\left(a, \dfrac{b}{M}, \dfrac{1}{N}\right)$.

c. If $f(x,y)$ and $f'_y(x,y)$ are continuous on R and $|f'_y(x,y)| \le K$ in rectangle R, then Lipschitz continuity criteria holds.

d. The unique solution of a Cauchy's problem can be found using the **Picard iterations**, based on constructing a sequences of functions $y_1,\ldots,y_n,\ldots$ following the formula

$$y_n = y_0 + \int_{x_0}^{x} f[t, y_{n-1}(t)]dt \ , \ n = 1,2,\ldots \ , \ y_0 = y(x_0)$$

which converges to a solution.

e. If, under the existence and uniqueness in b, $f(x,y)$ does not hold the Lipschitz continuity criteria, then (*) has a least one solution.

f. If $f(x,y)$ is continuous and has continuous partial derivatives to n-order, included, in the neighborhood of point (x_0, y_0), then the solution $y = y(x)$ of Cauchy's problem (*) is an $(n+1)$ times differentiable function.

Chapter 2: Linear n-th Order Differential Equations

2.1   Definition

a. The equation

$$y^{(n)} + P_1(x)y^{(n-1)}(x) + \ldots + P_{n-1}(x)y'(x) + P_n(x)y(x) = f(x) \ (*)$$

is a n-**th order linear equation**.

b. If $f(x) \equiv 0$ the equation is called a **homogeneous linear equation**.

c. If all coefficients $P_i(x)$ are continuous functions on $a \le x \le b$, then in the neighborhood of initial conditions

$$y(x_0) = y_0, y'(x_0) = y'_0, \ldots, y^{(n-1)}(x_0) = y_0^{(n-1)}, \quad x_0 \in (a, b) \quad (**)$$

the Cauchy's problem (*, **) has a unique solution.

2.2 Linear Operator $L[y]$

a. We denote $L[y] = y^{(n)} + P_1(x) y^{(n-1)} + \ldots + P_{n-1}(x) y' + P_n(x) y$

b. For every fixed c: $L[cy] = cL[y]$

c. $L[y_1 + y_2] = L[y_1] + L[y_2]$

d. If y_1 and y_2 are solutions of the equation $L[y] = 0$, then $y_1 + y_2$ is also a solution of that equation.

e. If ODE $L[y] = 0$ with real coefficients $P_i(x)$ has complex solution $y(x) = u(x) + v(x)i$, then $y_1 = u(x)$ and $y_2 = v(x)$ are real solutions of that equation.

f. If $y = y_1$ is a non-trivial solution of $L[y] = 0$ then, by substituting $y = y_1 z$ and $z' = u$, we get an $n-1$ order ODE for u.

2.3 Solutions of $L[y] = 0$

a. Functions $y_1(x), y_2(x), \ldots, y_n(x)$ are **linearly dependent (LD)** in interval $[a, b]$ if there exist n numbers $\alpha_1, \alpha_2, \ldots, \alpha_n$, not all of which are zeroes, such that $\sum_{i=1}^{n} \alpha_i y_i(x) \equiv 0$. Otherwise, functions $y_i(x)$ are **linearly independent (LI)**.

b. If functions $y_1(x), y_2(x), \ldots, y_n(x)$ are **LD** on $[a, b]$, then the determinant

$$w(x) = w[y_1 y_2 \cdots y_n] = \begin{vmatrix} y_1 & y_2 & \cdots & y_n \\ y'_1 & y'_2 & \cdots & y'_n \\ y''_1 & y''_2 & \cdots & y''_n \\ \vdots & \vdots & & \vdots \\ y_1^{(n-1)} & y_2^{(n-1)} & \cdots & y_n^{(n-1)} \end{vmatrix}$$

is identically equal to zero.

Determinant $w(x)$ is called **Wronskian**.

c. If **LI** functions $y_1, y_2, \ldots, y_n$ are solutions of $L[y] = 0$ with continuous coefficients $P_i(x)$ on $[a, b]$, then Wronskian $w[y_1 y_2 \cdots y_n]$ is zero nowhere on $[a, b]$.

d. Particular solutions of $L[y] = 0$, with a continuous coefficient, which are linear independent on $[a, b]$, form a basis of the solution space.

e. Any solution $y(x)$ of $L[y] = 0$ is a linear combination of n **LI** solutions $y_1(x), y_2(x), \ldots, y_n(x)$

$$y(x) = \sum_{i=1}^{n} a_i y_i(x)$$

2.4   Restoring Linear ODE

a. Let $y_1(x), y_2(x), \ldots, y_n(x)$ be n **LI** functions in $[a, b]$.

We construct an $n+1$ th order determinant

$$w[y_1, y_2, \ldots, y_n, y] = \begin{vmatrix} y_1 & y_2 & \cdots & y_n & y \\ y'_1 & y'_2 & \cdots & y'_n & y' \\ \vdots & \vdots & & \vdots & \vdots \\ y_1^{(n-1)} & y_2^{(n-1)} & \cdots & y_n^{(n-1)} & y^{(n-1)} \\ y_1^{(n)} & y_2^{(n)} & \cdots & y_n^{(n)} & y^{(n)} \end{vmatrix} = 0$$

Developing it by its last column, we get a linear ODE the solutions of which are the given functions.

b. **Abel's identity** formula: $w(x) = w(x_0)e^{-\int_{x_0}^{x} P_1(x)\,dx}$

2.5   Linear Homogeneous ODE with Constant Coefficients

Let

$$L[y] = a_0 y^{(n)} + a_1 y^{(n-1)} + \ldots + a_n y = 0$$

be the n-th-order linear ODE, when a_i are its constant real coefficients. Then, the solutions are in the form of $y = e^{kx}$. Substituting, we get **the characteristic equation**

$$a_0 k^n + a_1 k^{n-1} + \ldots + a_{n-1} k + a_n = 0$$

a. If the characteristic equation has n different real solutions $, k_1, k_2, \ldots, k_n$ then $e^{k_1 x}, e^{k_2 x}, \ldots, e^{k_n x}$ are n **LI** solutions forming a basis of solution space. The general solution is $y = C_1 e^{k_1 x} + C_2 e^{k_2 x} + \ldots + C_n e^{k_n x}$ when C_i are constants.

b. If the characteristic equation has complex solution $k = \alpha + i\beta$, then $k = \alpha - i\beta$ is also one of its solutions. It is written $e^{(\alpha \pm i\beta)x} = e^{\alpha x}(\cos\beta x \pm i\sin\beta x)$. Therefore, this ODE has two **LI** solutions, $e^{\alpha x}\cos\beta x$ and $e^{\alpha x}\sin\beta x$.

c. If a characteristic equation has a root k_1 with multiplicity p, then it fits p LI solutions $e^{k_1 x}, x e^{k_1 x}, \ldots, x^{p-1} e^{k_1 x}$.

d. If a characteristic equation has complex root $\alpha + i\beta$ with multiplicity q, then it has 2q real LI solutions:

$e^{\alpha x}\cos\beta x, x e^{\alpha x}\cos\beta x, \ldots, x^{q-1} e^{\alpha x}\cos\beta x$

$e^{\alpha x}\sin\beta x, x e^{\alpha x}\sin\beta x, \ldots, x^{q-1} e^{\alpha x}\sin\beta x$

2.6 Euler's Equation

The equation

$$a_0 x^n y^{(n)} + a_1 x^{n-1} y^{(n-1)} + \ldots + a_{n-1} xy' + a_n y = 0$$

when a_i are real constant coefficients, is called **Euler's equation.**

a. The substitution $x = e^t$ transforms Euler's equation into a constant coefficient linear ODE.

b. We look for a solution in the form of $y = x^r$. The result is a **characteristic equation** with respect to r :

$$a_0 r(r-1)\ldots(r-n+1) + a_1 r(r-1)\ldots(r-n+2) + \ldots + a_n = 0$$

c. For each real root r_i with multiplicity p we get p **LI** solutions

$$x^{r_i}, x^{r_i} \ln x, x^{r_i} \ln^2 x, \ldots, x^{r_i} \ln^{p-1} x$$

d. For each pair of complex roots $r = \alpha \pm i\beta$ with multiplicity q we get $2q$ **LI** solutions:

$$x^\alpha \cos(\beta \ln x), x^\alpha \ln x \cos(\beta \ln x), \ldots, x^\alpha \ln^{q-1} x \cos(\beta \ln x)$$

$$x^\alpha \sin(\beta \ln x), x^\alpha \ln x \sin(\beta \ln x), \ldots, x^\alpha \ln^{q-1} x \sin(\beta \ln x)$$

2.7 Non-homogeneous Linear ODE

a. The general solution of the ODE

$$L[y] = y^{(n)} + P_{n-1}(x) y^{(n-1)} + \ldots + P_0(x) y(x) = f(x) \quad (*)$$

is $y(x) = y_h(x) + y_p(x)$ when $y_h(x)$ is the general solution of the homogeneous equation and $y_p(x)$ is a particular solution of the non-homogeneous equation, respectively.

b. **The parameter variation method of finding a particular solution:**

If $y_h(x) = \sum_{i=1}^{n} C_i y_i(x)$ is a general solution of (*) in the homogeneous equation, we look for a solution of the following form:

$$y(x) = C_1(x)y_1 + C_2(x)y_2 + \ldots + C_n(x)y_n(x)$$

Substituting it in equation (*), we construct a linear equation system with respect to $C'_n(x), \ldots, C'_2(x), C'_1(x)$:

$$\begin{cases} C'_1(x)y_1 & + & C'_2(x)y_2 & + & \cdots & + & C'_n(x)y_n & = & 0 \\ C'_1(x)y'_1 & + & C'_2(x)y'_2 & + & \cdots & + & C'_n(x)y'_n & = & 0 \\ \vdots & & \vdots & & & & \vdots & & \\ C'_1(x)y_1^{(n-2)} & + & C'_2(x)y_2^{(n-2)} & + & \cdots & + & C'_n(x)y_n^{(n-2)} & = & 0 \\ C'_1(x)y_1^{(n-1)} & + & C'_2(x)y_2^{(n-1)} & + & \cdots & + & C'_n(x)y_n^{(n-1)} & = & f(x) \end{cases}$$

The determinant of the system is Wronskian $w[y_1 y_2 \ldots y_n] \neq 0$. Therefore, the system has a unique solution.

2.8   Non-homogeneous Linear ODE with Constant Coefficients

$$y^{(n)} + a_1 y^{(n-1)} + \ldots + a_{n-1}y' + a_n y = f(x) \ (*)$$

Characteristic equation fitting (*)

$$k^n + a_1 k^{n-1} + \ldots + a_{n-1}k + a_n = 0 \ (**)$$

If $f(x)$ is one of the following functions:

1. Polynomial $P_n(x)$

2. $P_n(x) \cdot e^{\alpha x}$

3. $e^{\alpha x} P_n(x) \sin \beta x$ or $e^{\alpha x} P_n(x) \cos \beta x$

Then, we can find a particular solution of (*) on the form presented in the table, where $P_n(x) = \sum\limits_{i=0}^{n} P_i x^n$ is a given polynomial, $A_n(x) = \sum\limits_{i=0}^{n} a_i x^i$ and $B_n(x) = \sum\limits_{i=0}^{n} b_i x^i$ are unknown polynomials, and α, β are real numbers:

$f(x)$	Solutions of (**)	Particular solution $y_n(x)$ in the form of
$P_n(x)$	$a_n \neq 0$	$A_n(x)$
	$k = 0$, solution with multiplicity s	$x^s \cdot A_n(x)$
$e^{\alpha x} P_n(x)$	α no solution	$e^{\alpha x} \cdot A_n(x)$
	α solution with multiplicity s	$x^s \cdot e^{\alpha x} \cdot A_n(x)$
$e^{\alpha x} P_{n_1}(x)\cos\beta x +$ $e^{\alpha x} P_{n_2}(x)\sin\beta x$	$\alpha \pm i\beta$, no solution	$e^{\alpha x}\left(A_n(x)\cos\beta x + B_n(x)\sin\beta x\right)$ $n = \max(n_1, n_2)$
	$\alpha \pm i\beta$ solution with multiplicity s	$x^s e^{\alpha x}\left(A_n(x)\cos\beta x + B_n(x)\sin\beta x\right)$ $n = \max(n_1, n_2)$

Chapter 3
Series Solutions of Second-Order ODE

3.1 Solutions Near an Ordinary Point

a. x_0 is an **ordinary point** of ODE

$$P(x)y'' + Q(x)y' + R(x)y = 0 \quad (*)$$

if functions $p(x) = \dfrac{Q(x)}{P(x)}$ and $q(x) = \dfrac{R(x)}{P(x)}$ can be developed to a power series about x_0 converging at $0 < |x - x_0| < \rho$.

b. **Theorem**: If $x = x_0$ is an ordinary point of $(*)$ then $(*)$ has a solution in the form of series $y(x) = \sum\limits_{n=0}^{\infty} a_n (x - x_0)^n$ with a convergence radius of at least ρ.

c. **Example**: **Airy** equation $y'' - xy = 0$

$x_0 = 1$ is a regular point. We look for the solution in the form $y(x) = \sum\limits_{n=0}^{\infty} a_n (x - 1)^n$. We differentiate $y(x)$ twice:

$$y'(x) = \sum_{n=1}^{\infty} a_n \cdot n(x-1)^{n-1} \quad , \quad y''(x) = \sum_{n=2}^{\infty} n(n-1)a_n(x-1)^{n-2}$$

Substituting in the equation, we get:

$$\sum_{n=2}^{\infty} n(n-1)a_n(x-1)^{n-2} - x\sum_{n=0}^{\infty} a_n(x-1)^n = 0$$

Writing $x = 1 + (x-1)$, shifting the indexes, we get

$$\sum_{n=0}^{\infty} (n+2)(n+1)a_{n+2}(x-1)^n - \sum_{n=0}^{\infty} a_n(x-1)^n - \sum_{n=1}^{\infty} a_{n-1}(x-1)^n = 0$$

Equating the coefficients of the same powers of $(x-1)$ to zero, we get recursion formula

$$2a_2 = a_0 \ , \ n = 0$$

$$(n+2)(n+1)a_{n+2} = a_n + a_{n-1} \quad , \quad n \geq 1$$

Therefore, the general solution is

$$y(x) = a_0 \left[1 + \frac{(x-1)^2}{2} + \frac{(x-1)^3}{6} + \dots \right] +$$

$$+ a_1 \left[(x-1) + \frac{(x-1)^3}{6} + \frac{(x-1)^4}{12} + \dots \right]$$

3.2   Chebyshev's Equation

$$(1-x)^2 y'' - xy' + \alpha^2 y = 0 \ , \ \alpha \in \mathbb{R}$$

a. $x_0 = 0$ is an **ordinary point**. The general solution is

$$y(x) = a_0 \left[1 + \sum_{n=1}^{\infty} \frac{[(2n-2)^2 - \alpha^2] \dots (4-\alpha^2)(-\alpha^2)}{(2n)!} \right] +$$

$$+ a_1 \left[x + \sum_{n=1}^{\infty} \frac{[(2n-1)^2 \cdot \alpha^2] \dots (3-\alpha^2)(1-\alpha^2)}{(2n+1)!} \right]$$

b.   If α is zero or an integer, then one solution is a polynomial. Such polynomials are called **Chebyshev's polynomials**.

c. $T_n = \cos[n(\arccos x)]$, $n = 0, 1, \dots$ are **Chebyshev's polynomials**.

d.   Recursion formula:

$$T_{n+1}(x) + T_{n-1}(x) = 2xT_n(x) \ , \ n = 1,2,\ldots, T_0(x) = 1 \ , \ T_1(x) = x$$

3.3 Legendre Polynomials

Legendre Equation: $(1-x^2)y'' - 2xy' + n(n+1)y = 0$

a. Its polynomial solutions $P_n(x)$ are Legendre Polynomials holding the recursion formula

$$(n+1)P_{n+1}(x) - nP_{n-1}(x) = (2n+1)xP_n(x), n =$$

$$= 1,2,\ldots, P_0(x) = 1, P_1(x) = x$$

b. **Rodrigues** formula: $P_n(x) = \dfrac{1}{2^n n!} \dfrac{d^n}{dx^n}(x^2 - 1)^n$

c. $\displaystyle\int_{-1}^{1} P_m(x) P_n(x)\,dx = \begin{cases} 0 & , \ m \neq 0 \\ \dfrac{2}{2n+1} & , \ m = n \end{cases}$

3.4 Solutions Near a Regular Singular Point

Point $x = x_0$ is a **regular singular point** of ODE

$$P(x)y'' + Q(x)y' + R(x)y = 0 \ (*).$$

If there exist finite limits

$$p_0 = \lim_{x \to x_0} \frac{(x-x_0)Q(x)}{P(x)} \ , \ q_0 = \lim_{x \to x_0} \frac{(x-x_0)^2 R(x)}{P(x)} \ ,$$

then the equation $r(r-1) + p_0 r + q_0 = 0 \ (**)$ is called the **characteristic equation** of $(*)$.

a. If the characteristic equation $(**)$ has two different real solutions, r_1, r_2, the difference of which, $r_1 - r_2$, is not an integer, then $(*)$ has two linearly independent **LI** solutions

$$y = \sum_{n=0}^{\infty} a_n (x - x_0)^{n+r_1} \quad , \quad y = \sum_{n=0}^{\infty} a_n (x - x_0)^{n+r_2}$$

b. If the characteristic equation (**) has two equal solutions $r_1 = r_2$, then (*) has two **LI** solutions

$$y_1(x) = \sum_{n=0}^{\infty} a_n (x - x_0)^{n+r_1} \quad ,$$

$$y_2 = y_1(x) \ln |x - x_0| + \sum_{n=1}^{\infty} b_n (x - x_0)^{n+r_1}$$

c. If $r_1 > r_2$ and $r_1 - r_2$ is an integer, then (*) has two **LI** solutions

$$y_1(x) = \sum_{n=0}^{\infty} a_n (x - x_0)^{n+r_1} \quad ,$$

$$y_2(x) = a y_1(x) \ln |x - x_0| + \sum_{n=0}^{\infty} b_n (x - x_0)^{n+r_2}$$

3.5   Bessel's Equation $x^2 y'' + x y' + (x^2 - v^2) y = 0$

a. The General solution, $y(x) = a_1 J_v(x) + a_2 J_{-v}(x)$ when

$$J_{\pm v}(x) = x^{\pm v} \sum_{m=0}^{\infty} \frac{(-1)^m x^{2m}}{2^{2n+v} m! \Gamma(m \pm v + 1)} \left(\frac{x}{2} \right)^{2m \pm v}$$

b. $J_{\frac{1}{2}}(x) = \sqrt{\dfrac{2}{\pi x}} \sin x \ , \ J_{-\frac{1}{2}}(x) = \sqrt{\dfrac{2}{\pi x}} \cos x$

c. $J_v(x)$ are **Bessel functions of the first kind**

d. For $v = n$ integer, the solutions of the equation are Bessel functions

$$J_n(x) = x^n \sum_{m=0}^{\infty} \frac{(-1)^m}{m!(n+m)!} x^{2m} \ , \ J_{-n}(x) = (-1)^n J_n(x)$$

e. Bessel equation $y''(x) + \dfrac{1}{x} y'(x) + y(x) = 0$ of order $\nu = 0$

has two solutions, one unbounded in the neighborhood of $x = 0$, and another is zero-order Bessel function

$$J_0(x) = \sum_{n=0}^{\infty} (-1)^n \frac{x^{2n}}{2^{2n}(n!)^2}.$$

Chapter 4
Systems of First-Order Linear Equations

4.1 Definition

a. The system

$$\begin{cases}
x'_1 = a_{11}(t)x_1 + a_{12}(t)x_2 + \ldots + a_{1n}(t)x_n + f_1(t) \\
x'_2 = a_{21}(t)x_1 + a_{22}(t)x_2 + \ldots + a_{2n}(t)x_n + f_2(t) \\
\vdots \qquad \vdots \qquad \vdots \qquad \qquad \vdots \qquad \vdots \\
x'_n = a_{n1}(t)x_1 + a_{n2}(t)x_2 + \ldots + a_{nn}(t)x_n + f_n(t)
\end{cases}$$

$$(*)$$

is a **non-homogeneous linear ODE** system.

b. If $f_i(t) \equiv 0$, $i = 1,2,\ldots,n$, then the system is **homogeneous.**

c. if functions $f_1(t),\ldots,f_n(t)$ and $a_{ij}(t)$, $(i,j = 1,2,\ldots,n)$ are continuous on (α,β), and $t_0 \in (\alpha,\beta)$, then system (*) has a unique solution holding the initial conditions.

$$x_1(t_0) = x_1^{(0)}, x_2(t_0) = x_2^{(0)},\ldots,x_n(t_0) = x_n^{(0)} \quad (**)$$

d. Vector form of a first-order ODE system

$$\mathbf{x'} = A\mathbf{x} + \mathbf{f} \ , \ \mathbf{x}(t_0) = \mathbf{x}_0$$

where

$$\mathbf{x} = \begin{pmatrix} x_1 \\ x_2 \\ \vdots \\ x_n \end{pmatrix} \ , \quad \mathbf{x'} = \begin{pmatrix} x'_1 \\ x'_2 \\ \vdots \\ x'_n \end{pmatrix} \ , \quad A = \begin{pmatrix} a_{11} & a_{12} & \cdots & a_{1n} \\ a_{21} & a_{22} & \cdots & a_{2n} \\ \vdots & \vdots & & \vdots \\ a_{n1} & a_{n2} & \cdots & a_{nn} \end{pmatrix} \ , \quad \mathbf{f} = \begin{pmatrix} f_1 \\ f_2 \\ \vdots \\ f_n \end{pmatrix}$$

4.2 Homogeneous System of ODE $\mathbf{x'} = A\mathbf{x}$

a. Set of Vectors $\mathbf{x}^{(1)}, \mathbf{x}^{(2)}, \ldots, \mathbf{x}^{(n)}$ is a basis of a solution space, if they are n **LI** solutions of system (*) on (α, β). In this case, any other solution is their linear combination.

b. The determinant $W = [\mathbf{x}^{(1)}, \mathbf{x}^{(2)}, \ldots, \mathbf{x}^{(n)}]$ is called **Wronskian**.

c. If $\mathbf{x}^{(1)}, \ldots, \mathbf{x}^{(n)}$ are n-solutions of (*) with continuous coefficients on (α, β) , then, in this interval, the Wronskian is identically equal to zero or never vanishes.

4.3 Homogeneous System of First-Order Linear Equations with Constant Coefficients

Let $\mathbf{x'} = A\mathbf{x}$ be linear ODE when A is constant matrix.

a. If matrix A has n different eigenvalues $\lambda_1, \lambda_2, \ldots, \lambda_n$ and $\mathbf{x}^{(1)}, \mathbf{x}^{(2)}, \ldots, \mathbf{x}^{(n)}$ eigenvectors corresponding them, then

$$\mathbf{x}^{(1)} e^{\lambda_1 t}, \mathbf{x}^{(2)} e^{\lambda_2 t}, \ldots, \mathbf{x}^{(n)} e^{\lambda_n t}$$

is a basis of the solution space.

b. If eigenvalue λ_k of matrix A has algebraic multiplicity k but its geometric multiplicity is smaller than the algebraic one, we construct k linearly independent vectors $\mathbf{c}^{(i)}$ in the following way

$$(A - \lambda_k I)\mathbf{c}^{(1)} = 0 \text{ , } (A - \lambda_k I)\mathbf{c}^{(i)} = \mathbf{c}^{(i-1)} \text{ , } i = 2,...,k$$

Vectors $\mathbf{c}^{(1)}, \mathbf{c}^{(2)},...,\mathbf{c}^{(k)}$ **are a set of LI** eigenvectors, **corresponding to** λ_k and

$$\mathbf{x}^{(1)}(t) = \mathbf{c}^{(1)} e^{\lambda_k t}$$
$$\mathbf{x}^{(2)}(t) = (\mathbf{c}^{(1)} t + \mathbf{c}^{(2)}) e^{\lambda_k t}$$
$$\vdots$$
$$\mathbf{x}^{(k)}(t) = \left(\mathbf{c}^{(1)} \frac{t^{k-1}}{(k-1)!} + \mathbf{c}^{(2)} \frac{t^{k-2}}{(k-2)!} + ... + \mathbf{c}^{(k)} \right) e^{\lambda_k t}$$

are k **LI** solutions of the system corresponding to λ_k.

4.4   Non-homogeneous System of First-Order Linear Equations

$$\mathbf{x}'(t) = A(t)\mathbf{x}(t) + \mathbf{f}(t) \quad (*)$$

Let

$$\psi(t) = [\mathbf{x}^{(1)}(t), \mathbf{x}^{(2)}(t),...,\mathbf{x}^{(n)}(t)]$$

be an $n \times n$ matrix, where the columns are solutions of the corresponding homogeneous system. The general solution of the homogeneous system is $\psi(t) \cdot \mathbf{c}$, when $\mathbf{c}$ is a constant vector.

Varying the parameter, we look for a particular solution of system (*) in the form of $\mathbf{x} = \psi(t) \cdot \mathbf{c}(t)$. Substituting in (*), we get equation system $\psi(t) \cdot \mathbf{c}'(t) = \mathbf{f}(x)$, the solutions of which are $\mathbf{c}(t) = \int \psi^{-1}(t) \mathbf{f}(t) dt + \mathbf{c}^{(1)}$. Therefore, the final solution is

$$\mathbf{x}(t) = \psi(t) \int \psi^{-1}(t) \mathbf{f}(t) dt + \psi(t) \cdot \mathbf{c}^{(1)}$$

when $c^{(1)}$ is an arbitrary vector.

Chapter 5
Sturm-Liouville Eigenvalue Problem

5.1   Definition

a. A second-order ODE

$$[r(x)y'(x)]' + [q(x) + \lambda p(x)]y(x) = 0 , \quad a < x < b \quad (*)$$

with boundary conditions

$$\begin{cases} \alpha_1 y(a) + \beta_1 y'(a) = 0 \\ \alpha_2 y(b) + \beta_2 y'(b) = 0 \end{cases} \quad (**)$$

when $\alpha_1, \alpha_2, \beta_1, \beta_2$ are given real numbers and λ is parameter, is called a **boundary problem** or a **Sturm-Liouville problem.**

b. The solution $y(x) \not\equiv 0$ of (*), (**) is called an **eigenfunction** of **Sturm-Liouville problem**.

c. A λ for which problem (*), (**) has a solution is called **an eigenvalue of the problem**.

d. Let functions $p(x)$, $q(x)$, $r(x)$ be real and continuous on $a \le x \le b$, $y_m(x)$ and $y_n(x)$ be eigenfunctions of a Sturm-Liouville problem corresponding to different eigenvalues λ_m, λ_n, respectively, when $y'_m(x)$, $y'_n(x)$ are continuous on $[a,b]$. Then, $y_m(x)$ and $y_n(x)$ are orthogonal with respect to weight function $p(x)$, that is,

$$\int_a^b p(x)y_m(x)y_n(x)dx = 0, \quad m \ne n$$

e. If, in addition to the conditions of paragraph d, $p(x) > 0$, $(p(x) < 0)$ for every $x \in [a, b]$, then there is a countable set of real eigenvalues $\lambda_1 < \lambda_2 < ... < \lambda_n < ...$

f. The orthogonal system $\{y_n(x)\}_{n=0}^{\infty}$ of all eigenfunctions of Sturm-Liouville problem (*) is complete (see X.13.6) in the space of piecewise continuous functions on (a, b).

g. If function $f(x)$ is a piecewise differentiable on $[a, b]$, then, for every x of the same interval, the expansion of $f(x)$ into a series of the eigenfunctions of Sturm-Liouville problem is

$$f(x) \sim \sum_{n=1}^{\infty} c_n y_n(x) \ , \ c_n = \int_a^b p(x) y_n(x) f(x) dx$$

This series converges to $f(x)$ in the points of continuity and to $\frac{1}{2}[f(x^+) + f(x^-)]$ in the points of discontinuity.

h. If $f(x)$ and $f'(x)$ are continuous and $f''(x)$ is piecewise continuous on $[a, b]$, then $f(x) = \sum_{n=1}^{\infty} c_n y_n(x)$, and the series is absolutely and uniformly convergent on $[a, b]$.

For example, if $P_n(x)$ are Legendre polynomials, then

$$f(x) = \sum_{n=0}^{\infty} a_n P_n(x), \quad a_n = \frac{2n+1}{2} \int_{-1}^{1} f(x) P_n(x) dx$$

5.2   Examples

a. The equation $y''(x) + \lambda^2 y(x) = 0$

Eigenfunctions	Eigenvalues	Boundary conditions
$y(0) = 0$ $y(\ell) = 0$	$\lambda_k = \dfrac{k\pi}{\ell}$, $k \in \mathbb{N}$	$y_k = \sin\dfrac{k\pi}{\ell}x$
$y'(0) = 0$ $y(\ell) = 0$	$\lambda_k = \dfrac{2k-1}{2\ell}\pi$, $k \in \mathbb{N}$	$y_k = \cos\dfrac{2k-1}{2\ell}\pi x$
$y(0) = 0$ $y'(\ell) = 0$	$\lambda_k = \dfrac{2k-1}{2\ell}\pi$, $k \in \mathbb{N}$	$y_2 = \sin\dfrac{2k-1}{2\ell}\pi x$
$y'(0) = 0$ $y'(\ell) = 0$	$\lambda_k = \dfrac{\pi k}{\ell}$, $k = 0,1,2,\ldots$	$y_k = \cos\dfrac{\pi k}{\ell}x$

b. Legendre equations (see 3.3), can also be written in the Sturm-Liouville form

$$[(1-x^2)y']' + \lambda y = 0 \ , \ \lambda = n(n+1)$$

In this case, $r(x) = 1-x^2$, $q(x) \equiv 0$, and $p(x) = 1$, $-1 \le x \le 1$. The fitting eigen functions are Legendre polynomial $P_n(x)$.

XII. Complex Functions

Chapter 1
Complex Numbers Sequence (see X.1)

1.1 ε - Neighborhood

a. A set of all points $U(z_0,\varepsilon)=\{z\in\mathbb{C}:|z-z_0|<\varepsilon\}$ is called a ε - **neighborhood** of complex number z_0.

b. z_0 is an **accumulation point** of set E if every neighborhood of z_0 has at least one point of E different from z_0.

1.2   Limit Point of a Sequence

a. The number $z_0\in\mathbb{C}$ is a limit of sequence $\{z_n\}$, $z_n\in\mathbb{C}, n\in\mathbb{N}$, if, for every $\varepsilon>0$ there exists an $N(\varepsilon)$, such than for every n greater than $N(\varepsilon)$, there holds $|z_n-z_0|<\varepsilon$. It is denoted $\lim_{n\to\infty}z_n=z_0$, or is said, sequence $\{z_n\}$ converges to z_0.

b. **Cauchy's Criteria**: $\{z_n\}$ converges to z_0 if and only if for every $\varepsilon>0$ there exists an $N(\varepsilon)$, such that for every $n>N(\varepsilon)$ and every integer p, $|z_{n+p}-z_n|<\varepsilon$.

Written in short:

$$\{z_n\}\to z_0 \Leftrightarrow \forall\varepsilon<0, \exists N(\varepsilon); \forall n>N(\varepsilon), \forall p\in\mathbb{N}:|z_{n+p}-z_n|<\varepsilon$$

c. The sequence of complex numbers $\{z_n\} = \{x_n + iy_n\}$ converges to the number $z_0 = x_0 + iy_0$ if and only if the sequences of real numbers $\{x_n\}$ and $\{y_n\}$ converge to x_0 and y_0, respectively.

d. If complex number sequences $\{z_n\}$ and $\{w_n\}$ converge, then sequences $\{z_n \pm w_n\}$, $\{z_n \cdot w_n\}$, and $\left\{\dfrac{z_n}{w_n}\right\}$ with $w_n \neq 0$, also converge and hold:

$$\lim_{n \to \infty}(z_n \pm w_n) = \lim_{n \to \infty} z_n \pm \lim_{n \to \infty} w_n$$

$$\lim_{n \to \infty} z_n \cdot w_n = \lim_{n \to \infty} z_n \cdot \lim_{n \to \infty} w_n$$

$$\lim_{n \to \infty} \frac{z_n}{w_n} = \frac{\lim\limits_{n \to \infty} z_n}{\lim\limits_{n \to \infty} w_n} \ , \ (\lim_{n \to \infty} w_n \neq 0).$$

1.3   Sets in $\mathbb{C}$

a. z_0 is an **inner point** of D if there exists an ε-neighborhood of z_0 which is entirely in D.

b. w is a **boundary point** of D if in every ε-neighborhood of w there are points of D and points not of D.

c. The set of all boundary points of D form a **boundary of set D**. It is denoted ∂D.

d. **Set D is open** if it only consists of its inner points.

e. **Set D is closed** if it includes all of its boundary points.

f. **Set D is closed** if, and only if, it contains all of its accumulation points.

g. **Set D is bounded** if there exists a circle of a finite radius containing it.

h. **Bolzano-Weierstrass theorem:** every infinite and bounded sequence of complex numbers has a subsequence converging to the limit.

1.4   Curves and Domains in $\mathbb{C}$

a. A **Jordan curve or** a **continuous curve** is the set of points in the complex plane $z = z(t) = x(t) + iy(t)$, $\alpha \le t \le \beta$ where $y(t)$, $x(t)$ are real continuous functions.

   If, in addition, for every two different values t $(t_1 \ne t_2)$, there are two different fitting points on that line $(z(t_1) \ne z(t_2))$ except probably $t_1 = \alpha$, $t_2 = \beta$, this is a **simple curve**.

b. A simple curve is a **smooth curve** if $x(t)$, $y(t)$ have continuous derivatives, which do not vanish simultaneously, that is $[x'(t)]^2 + [y'(t)]^2 \ne 0$, $t \in [\alpha, \beta]$.

c. A continuous curve is **piecewise smooth** if it consists of a finite number of smooth curves.

d. The **positive direction** of the curve is the parameter t increase direction.

e. Closed curve divides the plane into two domains, one not containing $z = \infty$ (an **inner domain** with respect to a closed continuous curve), and another, containing $z = \infty$ (an **outer region** with respect to the same curve). This curve is a boundary of each of these domains.

f. **A positive direction on a continuous curve** is such that while we move along the curve in that direction, the inner region is always in the left.

g. An inner region with respect to a continuous curve is called a **Simply Connected Domain**. Otherwise, it is a **multi-connected domain**.

h. A domain with a boundary consisting of $n+1$ closed curves $\Gamma_0, \Gamma_1, \ldots, \Gamma_n$ such that each of the curves $\Gamma_1, \ldots, \Gamma_n$ is outside the other curves, and all within Γ_0 is called an $(n+1)$ -**connected domain**.

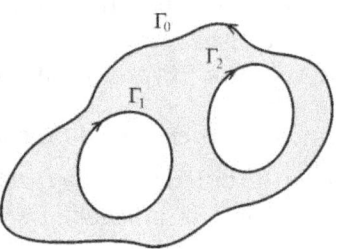

i. A **positive direction** on the boundary of $(n+1)$ connected domain is such that while we move along each of the boundary components, the domain is always on the left. In the illustration, the positive direction on Γ_0 is counterclockwise, while on Γ_2, Γ_1 it is clockwise.

j. Set D is connected if every two of its inner points can be connected with a continuous curve entirely contained in the set.

k. An open and connected set is called a **domain**.

Chapter 2
Complex Functions

2.1   Single-Valued and Multiple-Valued Functions

a. Let E be a set of complex numbers. If for every $z \in E$ there corresponds one or several complex numbers, following a given rule, then the function is said to be defined on E and we write $w = f(z)$, $z \in E$.

E is called the function's **domain of definition**.

b. If to each z, there corresponds one and only one value w, function $w = f(z)$ is said to be a **single-valued function**. If at least one z has several corresponding values, then the function is called a **multi-valued function**.

c. If $z = x + iy$, and we separate the real component from the imaginary component of $f(z)$, we get

$$w = u(x,y) + iv(x,y)$$

When $u(x,y) = \operatorname{Re} f(z)$, the real component of $f(z)$, $v(x,y) = \operatorname{Im} f(z)$ the imaginary component of $f(z)$ are real functions.

d. **Examples**:

1. Function $w = \dfrac{1}{z}$ is single-valued, defined on the full z-plane, except $z = 0$.

   If $z = x + iy$, then

$$w = \frac{x}{x^2 + y^2} - \frac{y}{x^2 + y^2}i$$

2. **Polynomial** $P_n(z) = \sum_{i=0}^{n} a_i z^i$ of complex variable z with a complex coefficient is a single-valued function defined on the z-plane.

3. **Rational function** $\dfrac{a_0 z^n + a_1 z^{n-1} + \ldots + a_n}{b_0 z^m + b_1 z^{m-1} + \ldots + b_m}$ is a single-valued function defined on the full z-plane, except the points vanishing the denominator.

4. Function $w = \sqrt[3]{3z-1}$ is defined on the full z-plane, and is three-valued function, since cube root has 3 values in a complex plane, for example:

$$w(0) = -1, \quad w(0) = \frac{1}{2} + \frac{\sqrt{3}}{2}i, \quad w(0) = \frac{1}{2} - \frac{\sqrt{3}}{2}i$$

5. A point that, when we move about it, has the initial value of a multi-valued function, and is transforming into another of its values, is called **a branch point**.

2.2   Limits of Complex Functions

a. Complex number w_0 is the limit of function $f(z)$, when $z \to z_0$, if for every $\varepsilon > 0$ there exists $\delta > 0$ such that every z holding $|z - z_0| < \delta$, there holds $|f(z) - w_0| < \varepsilon$.

It is denoted $\lim\limits_{z \to z_0} f(z) = w_0$.

b. $\lim\limits_{z \to \infty} f(z) = w_0$ if, for every $\varepsilon > 0$, there exists $R > 0$, such that for every $|z| > R$, then there holds $|f(z) - w_0| < \varepsilon$.

c. $\lim\limits_{z \to z_0} f(z) = \infty \left(\lim\limits_{z \to \infty} f(z) = \infty \right)$ if, for every $M > 0$, there exists a $\delta > 0$ $(R > 0)$, such that for every z holding $|z - z_0| < \delta$, $(|z| > R)$, then there holds $|f(z)| > M$.

d. Let $f(z) = u(x,y) + iv(x,y)$ and $w_0 = u_0 + iv_0$, $z_0 = x_0 + iy_0$. The limit $\lim\limits_{z \to z_0} f(z) = w_0$ exists if, and only if, the limits

$\lim\limits_{\substack{x \to x_0 \\ y \to y_0}} u(x,y) = u_0$, $\lim\limits_{\substack{x \to x_0 \\ y \to y_0}} v(x,y) = v_0$ exist.

2.3   Continuity

a. Function $f(z)$ is **continuous at point** z_0 if it is defined on the neighborhood of this point and $\lim\limits_{z \to z_0} f(z) = f(z_0)$.

b. Function $f(z)$ is **continuous in a domain** if it is continuous at every point of it.

c. If functions $f(z)$ and $g(z)$ are continuous on z_0, then also $f(z) \pm g(z)$, $f(z) \cdot g(z)$, $\dfrac{f(z)}{g(z)}$ when $g(z_0) \neq 0$, are continuous on z_0.

d. Function $f(z) = u(x,y) + iv(x,y)$ is continuous at point $z_0 = x_0 + iy_0$ if and only if real functions $u(x,y)$, $v(x,y)$ are continuous at (x_0, y_0).

e. Function $f(z)$ is **uniformly continuous** on D, if for every $\varepsilon > 0$ there exists $\delta(\varepsilon) > 0$ such that for every z_1, z_2 on D holding $|z_1 - z_2| < \delta$ there holds $|f(z_1) - f(z_2)| < \varepsilon$.

f. **Cantor's theorem**: A function continuous on bounded and closed domain $\bar{D}$ is uniformly continuous on it.

Chapter 3
Elementary Functions

3.1   Exponential Functions, Hyperbolic Functions

a. For every complex z,

$$e^z = 1 + z + \frac{z^2}{2!} + \ldots = \sum_{n=0}^{\infty} \frac{z^n}{n!} \ , \ (R = \infty) .$$

If z is real $(z = x)$, we get an expansion of real function e^x to a power series (see VI,4.7).

b. For every z_1 and z_2 there holds $e^{z_1} \cdot e^{z_2} = e^{z_1 + z_2}$.

c. The equation $e^z = 0$ has no (complex) solutions, that is, e^z has no zeroes.

d. **Hyperbolic sine**: $\sinh z = \dfrac{e^z - e^{-z}}{2}$ (see II.2.8, 2.9)

e. **Hyperbolic cosine**: $\cosh = \cosh z = \dfrac{e^z + e^{-z}}{2}$
    (see II.2.8, 2.5)

f. **Hyperbolic tangent**: $\tanh = \tan hz = \dfrac{\sin hz}{\cos hz} = \dfrac{e^z - e^{-z}}{e^z + e^{-z}}$

g. $\cos hz = 0$ has   an   infinite   number   of   solutions:
    $z = (k + \dfrac{1}{2}) \pi i$

h. $\sin hz = 0$ has an infinite number of solutions: $z = k\pi i$.

3.2 Trigonometric Functions

For every $|z| < \infty$:

a. $\sin z = z - \dfrac{z^3}{3!} + \dfrac{z^5}{5!} - \ldots = \displaystyle\sum_{n=0}^{\infty} \dfrac{(-1)^n}{(2n+1)!} z^{2n+1}$

b. $\cos z = 1 - \dfrac{z^2}{2!} + \dfrac{z^4}{4!} - \ldots = \displaystyle\sum_{n=0}^{\infty} \dfrac{(-1)^n}{(2n)!} z^{2n}$

c. The equation $\sin z = 0$ has real solutions only: $z = \pi k, k \in \mathbb{Z}$

d. The equation $\cos z = 0$, has real solutions only: $z = \dfrac{\pi}{2} + \pi k, k \in \mathbb{Z}$

3.3 Euler's Formula

a. For every complex z, $e^{iz} = \cos z + i \sin z$

b. $\sin z = \dfrac{1}{2i}(e^{iz} - e^{-iz})$, $\cos z = \dfrac{1}{2}(e^{iz} + e^{-iz})$

c. e^z is a periodic function, with a period of $2\pi i$ (see II. 1.4).

d. Functions $\cos z$ and $\sin z$ are periodic, with a period of 2π.

e. $\cos^2 z + \sin^2 z = 1$ for every z of $\mathbb{C}$.

3.4 Logarithmic Function

a. A (**natural**) **logarithm** of complex number z is a complex number w holding $e^w = z$. It is denoted $w = \operatorname{Ln} z$.

The natural logarithm of real number r is usually denoted as $\ln r$.

b. If $z = r(\cos\varphi + i\sin\varphi)$, then

$$\text{Ln}z = \ln r + i(\varphi + 2\pi k) \ , \ \ k = 0, \pm1, \pm2, \ldots$$

or $\qquad \text{Ln}z = \ln|z| + i\arg z$

c. $\text{Ln}z$ is a multivalued function.

d. For every z_2, z_1 different from zero, and for every real n there holds:

1. $\text{Ln}(z_1 \cdot z_2) = \text{Ln}z_1 + \text{Ln}z_2$

2. $\text{Ln}\dfrac{z_1}{z_2} = \text{Ln}z_1 - \text{Ln}z_2$

3. $\text{Ln}z^n = n\text{Ln}z$

4. $\text{Ln}\sqrt[n]{z} = \dfrac{1}{n}\text{Ln}z$

3.5   Inverse Trigonometric and Hyperbolic Functions

a. $\arcsin z$ is a complex number w holding $\sin w = z$

$$\arcsin z = -i\text{Ln}(iz + \sqrt{1-z^2})$$

b. $\arccos z$ is a complex number w holding $\cos w = z$

$$\arccos z = -i\text{Ln}(z + \sqrt{1-z^2})$$

c. $\arctan z = \dfrac{1}{2i}\text{Ln}\dfrac{1+iz}{1-iz}$

d. $\operatorname{ar\,sinh}z = \text{Ln}(z + \sqrt{1+z^2})$

e. $\operatorname{ar\,cosh}z = \text{Ln}(z + \sqrt{z^2-1})$

f. $\operatorname{ar\,tanh}z = \dfrac{1}{2}\text{Ln}\dfrac{1+z}{1-z}$

Chapter 4
Complex Function Derivative

4.1   Definition

a. If the single-valued function $f(z)$ has a finite limit $f'(z_0) = \lim\limits_{\Delta z \to 0} \dfrac{f(z_0 + \Delta z) - f(z_0)}{\Delta z}$, then $f(z)$ is said to be differentiable on z_0.

b. If function $f(z)$ is differentiable on z_0, it is continuous on that point.

4.2   Derivative Rules

If c is constant, and functions $f(z)$ and $g(z)$ are differentiable, then:

a. $(c)' = 0$

b. $(cf(z))' = cf'(z)$

c. $(f(z) \pm g(z))' = f'(z) \pm g'(z)$

d. $(f(z) \cdot g(z))' = f'(z) \cdot g(z) + f(z) \cdot g'(z)$

e. $\left(\dfrac{f(z)}{g(z)} \right)' = \dfrac{f'(z) \cdot g(z) - f(z) \cdot g'(z)}{(g(z))^2}$, $g(z) \neq 0$

f. A composite function derivative: If $f(z)$ is differentiable on z_0, and $g(w)$ is differentiable on $w_0 = f(z_0)$, then composite function $F(z) = g(f(z))$ is differentiable on z_0 and $F'(z_0) = g'(w_0) \cdot f'(z_0)$.

g. $(z^n)' = nz^{n-1}$

h. $(e^z)' = e^z$

i. $(\cos z)' = -\sin z$

j. $(\sin z)' = \cos z$

4.3 Cauchy-Riemann (CR) Criterion (Equations)

a. Function $f(z) = u(x,y) + iv(x,y)$ is differentiable at point
$z_0 = x_0 + iy_0$ if and only if $u(x,y)$ and $v(x,y)$ are
differentiable at (x_0, y_0), and

$$\frac{\partial u}{\partial x}(x_0, y_0) = \frac{\partial v}{\partial y}(x_0, y_0) \ , \ \frac{\partial u}{\partial y}(x_0, y_0) = -\frac{\partial v}{\partial x}(x_0, y_0)$$

In this case: $f'(z) = \dfrac{\partial u}{\partial x} + i\dfrac{\partial v}{\partial x} = \dfrac{\partial v}{\partial y} - i\dfrac{\partial u}{\partial y}$

b. CR criterion in polar coordinates: if

$$f(z) = u(r,\varphi) + iv(r,\varphi) \ , \ z = re^{i\varphi} = r(\cos\varphi + i\sin\varphi)$$

then, $f(z)$ is differentiable if, and only if,

$\dfrac{\partial u}{\partial r} = \dfrac{1}{r}\dfrac{\partial v}{\partial \varphi} \ , \ \dfrac{\partial u}{\partial \varphi} = -r\dfrac{\partial v}{\partial r}$ and its derivative is

$$f'(z) = \frac{r}{z}\left(\frac{\partial u}{\partial r} + i\frac{\partial v}{\partial r}\right) = \frac{1}{z}\left(\frac{\partial v}{\partial \varphi} - i\frac{\partial u}{\partial \varphi}\right)$$

4.4 Analytic Function

a. Single-valued function $f(z)$ is **analytic function** on z_0 if
there exists neighborhood $U(z_0, \varepsilon)$, such that $f(z)$ is
differentiable on every point of it.

b. If a function is analytic at a point, then it is continuous on
it.

c. Analytic function is sometimes called a **holomorphic function** or **regular function**.

d. An analytic function at every point of complex plane $\mathbb{C}$ is called an **entire function**.

e. A single-valued function is **analytic on domain** D if it is differentiable at every point of the domain.

f. A set of points where $f(z)$ is analytic has to be an open set. If $f(z)$ is said to be analytic in closed set $\bar{D}$, it means there exists open set D_1 containing $\bar{D}$ $(D_1 \supset \bar{D})$, where $f(z)$ is analytic.

g. If functions $f(z)$ and $g(z)$ are analytic on D, then functions $f(z) \pm g(z)$, $f(z) \cdot g(z)$ and $\dfrac{f(z)}{g(z)}$ when $g(z)$ does not zero at any point of D, are also analytic on D.

h. The composition of analytic functions is an analytic function.

i. A single-valued branch of a multivalued function is an analytic function in domain D if it is differentiable at any point of D.

4.5   Harmonic Functions

a. A two-variable real function $u(x,y)$ is **harmonic** in domain D if it is continuous and has continuous partial derivatives up to second order and satisfy the **Laplace equations** (see XIV.7).

$$u_{xx} + u_{yy} = 0$$

b. If function $f(z) = u(x,y) + iv(x,y)$ is analytic in domain D, then functions $u(x,y)$ and $v(x,y)$ are harmonic on D.

c. If $u(x,y)$ and $v(x,y)$ are harmonic on D and hold the C.R. criteria on D, then function $v(x,y)$ is called a **harmonic conjugate** function to $u(x,y)$ on D.

d. Function $f(z) = u(x,y) + iv(x,y)$ is analytic on D if and only if $v(x,y)$ is a harmonic conjugate to $u(x,y)$ on D.

e. For every function $u(x,y)$ harmonic on D, there exists harmonic function $v(x,y)$ conjugate to $u(x,y)$.

4.6   Conformal Mapping

a. The transforming of one complex plane on another one is called **conformal mapping or transformation** at z_0 if it preserves the magnitude and direction of the angles and expands constantly in all directions. In other words, a transformation is conformal if it transforms a small enough triangle the vertex of which is on z_0 to a small enough triangle similar to it.

b. $w = f(z)$ is conformal mapping in domain D if, and only if, function $f(z)$ is analytic on D and $f'(z) \neq 0$, $z \in D$.

c. **Riemann's theorem**: There exists analytic function $w = f(z)$ mapping simply connected domain D on simply connected domain Δ except in two cases:
D and/or Δ are full complex plane;
D and/or Δ are all full complex plane pierced in one point.

d. If function $w = f(z)$ is analytic in simply connected domain D and continuous on $\bar{D}$, objectively mapping the boundary ∂D of D on curve Γ in plane w and preserves the direction of Γ, then $f(z)$ conformly maps domain D to a domain bounded by curve Γ.

e. **Examples**:

1. **Displacement**: $w = z + b$

2. **Rotation**: $w = e^{\alpha i} z$ (α real constant). In this case $|w| = |z|$ and $\arg w = \alpha + \arg z$, that is, point z transforms to point w by a rotation of vector z around the origin at angle α.

3. **Extension**: $w = r \cdot z$, $r > 0$. In this case, $|w| = r|z|$ and $\arg w = \arg z$. Therefore, point z transforms to point w on line Oz at a r times the distance $|z|$. This mapping is **expansion** when $r > 1$, or contraction, when $0 < r < 1$.

4.7 Möbius Transformation

a. The transformation $w = \dfrac{az + b}{cz + d}$, $(ad - bc \neq 0)$

when a, b, c, d are constant complex numbers is called **Möbius transformation** or **bilinear transformation**.

b. Circles and straight lines in the complex plane are called **generalized circles**.

c. Möbius transformation maps generalized circles in plane z to generalized circles in plane w.

d. If z_1 and z_2 are points symmetrical about circle C in plane z, that is, they are on the ray originating from center O of circle C, and the product of distances $Oz_2 \cdot Oz_1$ equals to the square of the radius r of the circle, $Oz_2 \cdot Oz_1 = r^2$, then, after a Mobius mapping, their images w_1 and w_2 will be symmetrical with respect to circle L, the image of circle C.

e. There exists a unique Mobius transformation mapping three different points z_1, z_2, z_3, in plane z to three different points w_1, w_2, w_3 in plane w. This transformation is:

$$\frac{w - w_1}{w - w_3} : \frac{w_2 - w_1}{w_2 - w_3} = \frac{z - z_1}{z - z_3} : \frac{z_2 - z_1}{z_2 - z_3}$$

If one of the points equals ∞, then all the differences including that point are replaced with 1.

Chapter 5
Integrals of Complex Functions

5.1   Line Integral in the Complex Plane

Let $w = f(z) = u(x, y) + iv(x, y)$ be a continuous function in domain D. Suppose Γ is a continuous curve entirely in D, starting at z_0 and ending at z_1. Let us define integral over Γ directed from z_0 to z_1:

$$\int_\Gamma f(z)\,dz = \int_\Gamma u\,dx - v\,dy + i\int_\Gamma v\,dx + u\,dy$$

when the right side integrals are real line integrals (see IX, 9.1).

5.2   Properties of the Integral

a. $\int_\Gamma dz = z_1 - z_0$, when Γ is directed from z_0 to z_1.

b. $\int_\Gamma [f_1(z) + f_2(z)]dz = \int_\Gamma f_1(z)dz + \int_\Gamma f_2(z)dz$

c. $\int_\Gamma cf(z)dz = c\int_\Gamma f(z)dz$, c is a complex constant.

d. $\int_\Gamma f(z)dz = -\int_{\Gamma^-} f(z)dz$, when Γ^- is a curve in the opposite direction of Γ, that is, from z_1 to z_0.

e. If curve Γ consists of several curves $\Gamma_1, \Gamma_2, ..., \Gamma_n$, then

$$\int_\Gamma f(z)dz = \int_{\Gamma_1} f(z)dz + \int_{\Gamma_2} f(z)dz + ... + \int_{\Gamma_n} f(z)dz$$

f. If, along the line of Γ, function $f(z)$ is bounded, that is $|f(z)| \le M$ (M constant) and ℓ is the length of Γ (see formula in IX.5), then

$$\left|\int_\Gamma f(z)dz\right| \le M\ell$$

5.3   CauchyTheorem and its Applications

a. If function $f(z)$ is analytic in a simply-connected (multi-connected) closed and bounded domain $\bar{D}$ with boundary Γ, then

$$\int_\Gamma f(z)dz = 0$$

b. If function $f(z)$ is analytic in multi-connected closed domain $\bar{D}$, then the integral along the interior boundary paths equals the sum of integrals by the inner boundary paths of domain $\bar{D}$, when all boundary paths are directed counterclockwise.

c. Let $f(z)$ be analytic function in a simply-connected domain D. Let z_0 and z be points belonging to domain D. Then, the integral $\int_{\Gamma} f(z)\,dz$ is independent of the path Γ connecting z_0 and z, directed from z_0 to z, which is entirely in D.

d. If function $f(z)$ is analytic in simply-connected domain D, then function $F(z) = \int_{z_0}^{z} f(t)\,dt$, is also analytic in D, and $F'(z) = f(z)$. $F(z)$ is called the anti-derivative of $f(z)$.

e. **Newton-Leibniz formula**: If $F(z)$ is the **anti derivative** of $f(z)$, then

$$\int_{z_1}^{z_2} f(z)\,dz = F(z_2) - F(z_1)$$

5.4   Cauchy Integral Formula and its Applications

a. If function $f(z)$ is analytic in simply-connected domain D and Γ is a continuous closed curve contained in D, then, for every inner point z_0 of Γ there holds

$$f(z_0) = \frac{1}{2\pi i} \int_{\Gamma} \frac{f(z)}{z - z_0}\, dz$$

b. If function $f(z)$ is analytic in closed domain $\bar{D}$ with boundary Γ, then it is infinitely differentiable at every point z_0 in D and its n-th derivative equals to

$$f^{(n)}(z_0) = \frac{n!}{2\pi i} \int_\Gamma \frac{f(z)}{(z-z_0)^{n+1}} \, dz \ , \ n = 1, 2, \ldots$$

c. **Morera's theorem**: If function $f(z)$ is continuous on D and $\int_\Gamma f(z)\,dz = 0$ for every closed path Γ in D, then $f(z)$ is analytic in D.

d. **Maximum (absolute value) module principle**: If non-constant function $f(z)$ is analytic in bounded domain D and continuous in $\bar{D}$, then the module of $f(z)$ attains its maximum value on the boundary of D.

e. **Liouville's theorem**: If function $f(z)$ is analytic over the whole (entire) plane and bounded over it, that is, there exists a positive number M such that for every z there holds $|f(z)| < M$, then $f(z)$ is a constant function, that is, $f(z) \equiv c$ for every z.

f. **Fundamental theorem of algebra**: Every equation in the form of

$$P_n(z) = a_0 z^n + a_1 z^{n-1} + \ldots + a_n = 0 \ , \ (n \geq 1)$$

with complex coefficients has at last one root, that is, there is a least one complex z_0 such that $P_n(z_0) = 0$.

Chapter 6
Taylor and Laurent Series

6.1   Regular and singular points of complex function

a. $z = a$ is a **regular point** of function $f(z)$ if the function is analytic in that point.

b. $z = a$ is a **singular point** of function $f(z)$ if the function is not analytic in any neighborhood of a.

c.   If a function is analytic in domain D than each point of D is regular.

6.2   Expansion of Taylor series

a.   Let $f(z)$ be analytic on D. Let $z = a$ be any point on D and C be a circle with its center at $z = a$ and radius r contained in D. Then, there exists a unique power series, with a radius of convergence of at least r converging to $f(z)$, which is

$$f(z) = \sum_{n=0}^{\infty} c_n (z-a)^n = c_0 + c_1 (z-a) + c_2 (z-a)^2 + \ldots \quad (*)$$

where $c_n = \dfrac{1}{2\pi i} \displaystyle\int_C \dfrac{f(t)\,dt}{(t-a)^{n+1}} = \dfrac{1}{n!} f^{(n)}(a)$, $|c_n| \leq \dfrac{M}{r^n}$, M is the maximum value of $|f(z)|$ on circle $|z-a| = r$.

(*) is a Taylor series of $f(z)$ at a.

In the case of $a = 0$, (*) is **Maclaurin** series.

b. The **radius of convergence** of Taylor series at a, converging to function $f(z)$, equals to the shortest distance between point $z = a$ and the **singular** point of $f(z)$ closest to $z = a$.

6.3   Zeroes of Analytical Function

a. Point $z = a$ is called a **zero** of $f(z)$ if $f(a) = 0$.

b. $z = a$ is a **zero of order** n of function $f(z)$ if

$$f(a) = 0, \ f'(a) = 0, \ldots, f^{(n-1)}(a) = 0, \ f^{(n)}(a) \neq 0$$

c. $z = a$ is a zero of order n of function $f(z)$ if and only if $f(z)$ can be represented the following way:

$$f(z) = c_n (z-a)^n + c_{n+1}(z-a)^{n+1} + \ldots = (z-a)^n \varphi(z)$$

where $\varphi(z)$ is an analytic function on $z = a$ and $\varphi(a) \neq 0$.

d. $z = a$ is called isolated zero of function $f(z)$ if there exists a neighborhood of a which does not contain further zeroes of $f(z)$.

e. The zeroes of a non-zero analytic function are isolated.

6.4   Laurent Series

a. A **two sided series** containing positive and negative powers of $z - a$

$$\sum_{n=-\infty}^{\infty} c_n (z-a)^n = \sum_{n=0}^{\infty} c_n (z-a)^n + \sum_{n=1}^{\infty} \frac{c_{-n}}{(z-a)^n}$$

is called a **Laurent series**.

The first series on the right side is a power series converging in circle $|z-a| < R$. The second series in the same side, after substituting $t = \dfrac{1}{z-a}$, turns into a power series $\sum_{n=1}^{\infty} c_{-n} t^n$ converging in circle $|t| < R_1$. Therefore, the second series converges at $|z-a| > \dfrac{1}{R_1} = r$. If, in addition, $r < R$, then the domain of convergence of a Laurent series is the common domain of convergence of the two series $r < |z-a| < R$. Otherwise, there is no point where the series converges.

b. If $f(z)$ is analytic in ring $r < |z-a| < R$, then there exists a unique Laurent series converging to $f(z)$ in that ring, with the coefficients

$$c_n = \frac{1}{2\pi i} \int_{\gamma} \frac{f(z)dz}{(z-a)^{n+1}} \ , \ n = 0, \pm 1, \ldots, \ \gamma = \{z : |z-a| = \rho, r < \rho < R\}$$

Chapter 7
Isolated Singular Point

7.1 Definitions

a. Singular point $z = a$ of $f(z)$ is an **isolated singular point** of $f(z)$ if there exists a ring $0 < |z-a| < R$ (small enough), where $f(z)$ is analytic, and therefore $f(z)$ can be expanded to Laurent series:

$$f(z) = \sum_{n=0}^{\infty} c_n (z-a)^n + \sum_{n=1}^{\infty} \frac{c_{-n}}{(z-a)^n} \qquad (*)$$

b. The second sum of (*) is called the **principal part** of $f(z)$.

c. In (*), one can notice three distinct cases:

1. A Laurent series not including its principal part, that is $c_{-n} = 0$, $n \in \mathbb{N}$

2. A principal part with a finite number of terms.

3. A principal part with an infinite number of terms.

7.2   Removable Singular Point

a. If, in an expansion of Laurent series (*) the coefficients are $c_{-n} = 0, n = 1,2,...$, then point $z = a$ is a **removable singular point**.

b. Point $z = a$ is a removable singular point of function $f(z)$ if and only if there exists a finite limit $\lim\limits_{z \to a} f(z)$.

c. **Example**: $z = 0$ is a removable singular point of function $f(z) = \dfrac{\sin z}{z}$.

7.3   Pole

a. If the principal part of a Laurent series contains a finite number of terms, $c_{-n} = 0, n > m$ and $c_{-m} \neq 0$, then

$$f(z) = \sum_{n=0}^{\infty} c_n (z-a)^n + \frac{c_{-1}}{z-a} + \frac{c_{-2}}{(z-a)^2} + ... + \frac{c_{-m}}{(z-a)^m} .$$

In this case, $z = a$ is called a **pole of order** m of $f(z)$.

b. If $m = 1$ then $z = a$ is a **simple pole** or just a pole of $f(z)$.

c. $z = a$ is a pole of order m if and only if $f(z)$ can be represented in the form of $f(z) = \dfrac{\psi(z)}{(z-a)^m}$ where $\psi(z)$ is analytic in the neighborhood of $z = a$ and $\psi(a) \neq 0$.

d. $z = a$ is a pole of order m of function $f(z)$ if and only if $z = a$ is a zero of order m of function $g(z) = \dfrac{1}{f(z)}$.

7.4 Essential Singularity

a. If the principal part of an expansion of function $f(z)$ to a Laurent series has an infinite number of terms, then $z = a$ is called **essential singular point** of $f(z)$.

b. **Picard theorem**: An analytic function in a perforated neighborhood of an isolated essential singular point attains (an infinite number of times) any finite value except, perhaps, one singe value. In other words, if $z = a$ is an isolated essential singular point of function $f(z)$ then, for every finite c, except perhaps one single value, equation $f(z) = c$ has an infinite number of solutions tending to a.

Chapter 8
Behavior of Analytical Functions at Infinity

a. Point z is said to belong to $R > 0$ – neighborhood of ∞ if $\dfrac{1}{z}$ belongs to $\varepsilon = \dfrac{1}{R}$ - neighborhood of $z = 0$.

b. Function $f(z)$ is analytic in the neighborhood of $z = \infty$, if function $f(\dfrac{1}{z})$ is analytic in the neighborhood of $z = 0$.

c. $z = \infty$ is a removable singular point, a pole or an essential singular point of function $f(z)$ if $z = 0$ is a removable singular point, a pole or an essential singular point of function $f(\frac{1}{z})$, respectively.

Chapter 9
Residue and its Applications

9.1   Definition

a. Coefficient c_{-1} in an expansion of $f(z)$ into a Laurent series in ring $0 < |z-a| < R$ is called a **residue** of $f(z)$ on $z = a$. It is denoted $c_{-1} = \text{res}(f, a)$.

b. $c_{-1} = \dfrac{1}{2\pi i} \int_{\Gamma} f(z) dz$ when Γ is a closed curve surrounding $z = a$ and is entirely in that ring.

9.2   Calculating Residues

a. If $z = a$ is a simple pole of $f(z)$, then

$$c_{-1} = \lim_{z \to a} (z-a) \cdot f(z)$$

b. If point $z = a$ is a pole of order m of $f(z)$, then

$$\text{res}(f, a) = \frac{1}{(m-1)!} \lim_{z \to a} \left[(z-a)^m f(z) \right]^{(m-1)}.$$

c. **Residue theorem**: Let $f(z)$ be an analytic function in domain D except a finite number of singular points

$a_1, a_2, \ldots, a_n$, then, for every closed curve Γ surrounding these points, and is in D, there holds

$$\int_\Gamma f(z)dz = 2\pi i \sum_{j=1}^n res(f, a_j)$$

Chapter 10
Poles and Zeroes of Meromorphic Functions

a. $f(z)$ is called a **meromorphic function** in domain D if all of its singular points in D are isolated, no necessarily simple poles.

b. If function $f(z)$ is meromorphic in domain D with boundary Γ and $f(z)$ has no zeroes or poles on Γ, then

$$\frac{1}{2\pi i} \int_\Gamma \frac{f'(z)}{f(z)} \, dz = N - P$$

where N is the number of zeroes of $f(z)$ on D, and P is the number of poles of $f(z)$ on D (a zero or a pole of order n is counted n times).

c. **Rouche theorem**: If functions $f(z)$ and $g(z)$ are analytic in domain $\bar{D}$ with boundary Γ, and on Γ, there holds $f(z) \neq 0$, $|g(z)| < |f(z)|$, then functions $f(z)$ and $f(z) + g(z)$ have the same number of zeroes in D.

XIII. Fourier Series and Integral Transforms

Chapter 1
Trigonometric Fourier Series

1.1   Basic Concepts

a. Let E be the **inner product space** of piecewise continuous functions defined on $[-\pi, \pi)$ and with values in $\mathbb{C}$:

$$f \in E \Rightarrow f : [-\pi, \pi) \to \mathbb{C}$$

$f(x)$ can be extended periodically on all $\mathbb{R}$ such that $f(x + 2\pi) = f(x)$.

b. For $f, g \in E$ the **inner product** $\langle f, g \rangle = \dfrac{1}{\pi} \int\limits_{-\pi}^{\pi} f(x) \overline{g(x)} \, dx$.

c. The **norm** of $f(x)$ is $\| f(x) \| = \left(\dfrac{1}{\pi} \int\limits_{-\pi}^{\pi} | f(x) |^2 \, dx \right)^{\!\! \frac{1}{2}}$.

d. The sequence of functions

$$\frac{1}{\sqrt{2}}, \sin x, \cos x, \sin 2x, \cos 2x, \sin 3x, \cos 3x, \ldots$$

is an **orthonormal system** on E (see X.13.3).

1.2 Fourier Series

a. The series $\dfrac{a_0}{2} + \sum\limits_{n=1}^{\infty} [a_n \cos nx + b_n \sin nx]$

when $a_n = \dfrac{1}{\pi} \int\limits_{-\pi}^{\pi} f(x)\cos nx\,dx$, $n = 0,1,2,\ldots$

$\qquad\qquad b_n = \dfrac{1}{\pi} \int\limits_{-\pi}^{\pi} f(x)\sin nx\,dx$, $n = 1,2,\ldots$

is a **Fourier series** corresponding to $f(x) \in E$. a_n and b_n are the **Fourier coefficients** of $f(x)$.

b. A corresponding Fourier series to $f(x)$ defines periodic function $g(x)$ on $\mathbb{R}$, with a period 2π, that is, $\forall x \in \mathbb{R}$, $g(x+2\pi) = g(x)$ and $g(x) = f(x)$, $-\pi < x < \pi$, except, perhaps, the points of discontinuity of $f(x)$.

c. **Example:** Given $f(x) = x + \pi$, $-\pi < x < \pi$, Fourier coefficients are

$$a_0 = \frac{1}{\pi} \int\limits_{-\pi}^{\pi} (x+\pi)dx = 2\pi \quad , \quad a_n = \frac{1}{\pi} \int\limits_{-\pi}^{\pi} (x+\pi)\cos nx\,dx = 0$$

$$b_n = \frac{1}{\pi} \int\limits_{-\pi}^{\pi} (x+\pi)\sin nx\,dx = -\frac{2}{n}\cos n\pi = \frac{(-1)^{n+1}2}{n}$$

and the Fourier series is $f(x) \sim \pi + 2\sum\limits_{n=1}^{\infty} \dfrac{(-1)^{n+1}}{n}\sin nx$.

The following illustration shows functions $f(x)$ and $S_3 = \pi + 2\sin x - \sin 2x + \dfrac{2}{3}\sin 3x$, which is a partial sum of $f(x)$'s Fourier series:

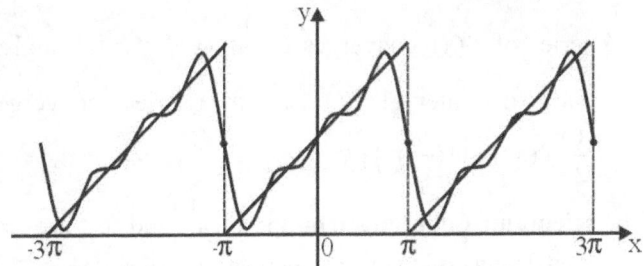

d. **Uniqueness theorem**: Let $f_1, f_2 \in E$. If the Fourier series corresponding to f_1 is equal to that of f_2, then $f_1(x) = f_2(x)$ in each $x \in [-\pi, \pi]$ except, perhaps, a finite number of points.

e. For every **trigonometric polynomial**

$$P_m(x) = \frac{A_0}{2} + \sum_{n=1}^{m} (A_n \cos nx + B_n \sin nx),$$ the expression

$\frac{1}{\pi} \int_{-\pi}^{\pi} |f(x) - P_m(x)|^2 \, dx$ attains its minimum when A_n and B_n are the Fourier coefficients of $f(x)$ on $[-\pi, \pi]$.

f. If $f(x)$ is an **even function**, then

$$f(x) \sim \frac{a_0}{2} + \sum_{n=1}^{\infty} a_n \cos nx \;, \quad a_n = \frac{2}{\pi} \int_0^{\pi} f(x) \cos nx dx$$

g. If $f(x)$ is an **odd function**, then

$$f(x) \sim \sum_{n=1}^{\infty} b_n \sin nx \;, \quad b_n = \frac{2}{\pi} \int_0^{\pi} f(x) \sin nx dx$$

1.3   Convergence of Fourier Series

a. **Dirichlet theorem**: If $f(x) \in E$ and $f(x)$ has one-sided derivatives (see IV.2.3, 4.6), in every point on $[\pi, -\pi]$, then, for every $x \in (-\pi, \pi)$, the corresponding Fourier series of $f(x)$ converges to $\dfrac{1}{2}[f(x^-)+f(x^+)]$, and at the ends of interval $x=\pm\pi$, the series converges to $\dfrac{1}{2}\Big[f(\pi^-)+f\big((-\pi)^+\big)\Big]$.

b. **Riemann-Lebesgue lemma**: If a_n and b_n are Fourier coefficients of $f(x)\in E$, then $\lim\limits_{x\to\infty} a_n = \lim\limits_{x\to\infty} b_n = 0$.

c. **Bessel's inequality**: If a_n and b_n are Fourier coefficients of $f(x)\in E$, then, $\dfrac{|a_0|^2}{2}+\sum\limits_{n=1}^{\infty}(|a_n|^2+|b_n|^2)\le\|f\|^2$

d. **Parseval's identity**: For every $f(x)\in E$, there holds

$$\frac{1}{\pi}\int_{-\pi}^{\pi}|f(x)|^2\,dx = \frac{|a_0|^2}{2}+\sum_{n=1}^{\infty}\Big(|a_n|^2+|b_n|^2\Big)$$

when a_n and b_n are Fourier coefficients of $f(x)$.

e. **Generalized Parseval's identity**: For every $f,g\in E$,

$$\frac{1}{\pi}\int_{-\pi}^{\pi}f(x)\overline{g(x)}\,dx = \frac{a_0\overline{c_0}}{2}+\sum_{n=1}^{\infty}(a_n\overline{c_n}+b_n\overline{d_n})$$

when a_n,b_n are Fourier coefficients of $f(x)$, and c_n,d_n are Fourier coefficients of $g(x)$.

1.4   Uniform Convergence

a. If function $f(x)$ is continuous on $[-\pi,\pi]$, $f(-\pi)=f(\pi)$, and $f'(x)\in E$, then Fourier series corresponding to $f(x)$ uniformly converges to $f(x)$ on every interval of $\mathbb{R}$.

b. Let $f,f'\in E$ and let $-\pi<d_1<d_2<\ldots<d_n<\pi$ be all the jump points of $f(x)$ on $(-\pi,\pi)$. If $[a,b]$ is a subinterval of $(-\pi,\pi)$ containing no point d_k, then Fourier series of f uniformly converges to f on $[a,b]$.

c. If $f(x) \in E$ and $f(x) \sim \dfrac{a_0}{2} + \sum\limits_{n=1}^{\infty} (a_n \cos nx + b_n \sin nx)$, then for every $x \in [-\pi, \pi]$ there holds

$$\int\limits_{-\pi}^{x} f(t)\,dt = \frac{a_0}{2}(x+\pi) + \sum_{n=1}^{\infty} \left(\frac{a_n}{n} \sin nx - \frac{b_n}{n}(\cos nx - \cos n\pi) \right)$$

and the series uniformly converges on $[-\pi, \pi]$.

d. If $f(x)$ is continuous on $[-\pi, \pi]$, $f(-\pi) = f(\pi)$, and $f'(x)$ is piecewise continuous on $[-\pi, \pi]$, then, the differentiation term by term of Fourier series corresponding to $f(x)$, results in Fourier series of $f'(x)$:

$$f'(x) \sim \sum_{n=1}^{\infty} (-na_n \sin nx + nb_n \cos nx)$$

1.5 Even and Odd Extension

a. If $f(x)$ is defined on $[0, \pi]$, then function

$$\hat{f}(x) = \begin{cases} f(x) , & 0 \le x \le \pi \\ f(-x) , & -\pi \le x < 0 \end{cases}$$

is an **even extension** of $f(x)$ to $[-\pi, \pi]$.

b. $\hat{f}(x) \sim \dfrac{a_0}{2} + \sum\limits_{n=1}^{\infty} a_n \cos nx$, $a_n = \dfrac{2}{\pi}\int\limits_{0}^{\pi} f(x)\cos nx\,dx$, $n = 0, 1, \ldots$

c. If $f(x)$ is defined on $[0, \pi]$, then function

$$\tilde{f}(x) = \begin{cases} f(x) , & 0 < x \le \pi \\ 0 , & x = 0 \\ -f(-x) , & -\pi \le x < 0 \end{cases}$$

is an **odd extension** of $f(x)$ to $[-\pi, \pi]$.

$$\tilde{f}(x) \sim \sum_{n=1}^{\infty} b_n \sin nx , \quad b_n = \frac{2}{\pi}\int\limits_{0}^{\pi} f(x)\sin nx\,dx, \quad n = 1, 2, \ldots$$

1.6 Fourier Series in Arbitrary Intervals

a. If $f(x)$ is defined on $[-T,T]$ and is piecewise continuous, then Fourier series of $f(x)$ on $[-T,T]$ is

$$f(x) \sim \frac{a_0}{2} + \sum_{n=1}^{\infty} \left(a_n \cos\frac{n\pi x}{T} + b_n \sin\frac{n\pi x}{T} \right)$$

when

$$a_n = \frac{1}{T} \int_{-T}^{T} f(x)\cos\frac{n\pi x}{T} dx \ , \ n = 0,1,\ldots$$

$$b_n = \frac{1}{T} \int_{-T}^{T} f(x)\sin\frac{n\pi x}{T} dx \ , \ n = 1,2,\ldots$$

b. The sequence of functions

$$\left\{ \frac{1}{\sqrt{2}} \ , \ \cos\frac{2\pi nx}{b-a} \ , \ \sin\frac{2\pi nx}{b-a} \ ,\ldots \right\}_{n=1}^{\infty} \quad \text{is an orthonormal}$$

system in the space of piecewise continuous functions on $[a,b]$ with inner product $\langle f,g \rangle = \dfrac{2}{b-a} \int_{a}^{b} f(x)\overline{g(x)} dx$.

c. If $f(x)$ is piecewise continuous on $[a,b]$, then the series

$$\frac{a_0}{2} + \sum_{n=1}^{\infty} \left(a_n \cos\frac{2n\pi x}{b-a} + b_n \sin\frac{2n\pi x}{b-a} \right)$$

$$a_n = \frac{2}{b-a} \int_{a}^{b} f(x)\cos\frac{2n\pi x}{b-a} dx \ , \ n = 0,1,2,\ldots$$

$$b_n = \frac{2}{b-a} \int_{a}^{b} f(x)\sin\frac{2n\pi x}{b-a} dx \ , \ n = 1,2,\ldots$$

is Fourier series corresponding to $f(x)$ on $[a,b]$.

1.7   Complex Fourier Series

a. The set of functions $\{e^{inx}\}_{n=-\infty}^{\infty}$ is an orthonormal system with respect to the inner product

$$\langle f,g \rangle = \frac{1}{2\pi} \int_{-\pi}^{\pi} f(x)\overline{g(x)}\,dx \ , \ f,g \in E .$$

b. Series $f(x) \sim \sum_{n=-\infty}^{\infty} c_n e^{inx}$,

when $c_n = \frac{1}{2\pi} \int_{-\pi}^{\pi} f(x)e^{-inx}dx$, $n = 0, \pm 1, \pm 2, \ldots$

is called **complex Fourier series** corresponding to $f(x)$.

c. If a_n and b_n are Fournier coefficients of trigonometric series $\frac{a_0}{2} + \sum_{n=1}^{\infty} a_n \cos nx + b_n \sin nx$, then

$$c_0 = \frac{a_0}{2}, \ c_{-n} = \frac{1}{2}(a_n + ib_n), \ c_n = \frac{1}{2}(a_n - ib_n) \ , \ n = 1, 2, \ldots$$

Chapter 2
Fourier Integral and Fourier Transform

Let $G(\mathbb{R})$ be set of all piecewise continuous functions defined on all $\mathbb{R}$ with values in $\mathbb{C}$, and absolutely integrable, that is, $\int_{-\infty}^{\infty} |f(x)|\,dx < \infty$.

2.1 Fourier Integral

a. If $f(x) \in G(\mathbb{R})$ and $f'(x) \in E(\mathbb{R})$, then, for all points of continuity of $f(x)$, then holds:

$$f(x) = \int_0^\infty (a_\omega \cos \omega x + b_\omega \sin \omega x) d\omega \qquad (*)$$

where $a_\omega = \dfrac{1}{\pi} \int_{-\infty}^\infty f(t) \cos \omega t \, dt$, $b_\omega = \dfrac{1}{\pi} \int_{-\infty}^\infty f(t) \sin \omega t \, dt$

Integral (*) is called the **Fourier integral** of $f(x)$.

b. Fourier integral (*) can be written this way:

$$f(x) = \frac{1}{\pi} \int_0^\infty d\omega \int_{-\infty}^\infty f(t) \cos \omega (t - x) dt$$

c. The **complex form** of Fourier integral is

$$f(x) = \frac{1}{2\pi} \int_{-\infty}^\infty d\omega \int_{-\infty}^\infty f(t) e^{-i\omega(t-x)} dt$$

2.2 Fourier Transform (Definitions)

a. $F(\omega) = \mathbb{F}[f](\omega) = \dfrac{1}{\sqrt{2\pi}} \int_{-\infty}^\infty f(x) e^{-i\omega x} dx$

is called the **Fourier transform** of $f(x)$.

b. If $f \in G(\mathbb{R})$, then $\mathbb{F}[f](\omega)$ is defined for every $\omega \in \mathbb{R}$, continuous on $\mathbb{R}$ and $\lim_{\omega \to \pm\infty} F(\omega) = 0$.

c. $f(x) = \mathbb{F}^{-1}[F](x) = \dfrac{1}{\sqrt{2\pi}} \int_{-\infty}^\infty F(\omega) e^{i\omega x} d\omega$

is called an **inverse Fourier transform**.

d. **Theorem:** If $f \in G(\mathbb{R})$, then for every $x \in \mathbb{R}$, then $f(x)$ has one-sided derivatives and holds

$$\frac{f(x^-)+f(x^+)}{2} = \lim_{M\to\infty}\frac{1}{\sqrt{2\pi}}\int_{-M}^{M}\mathbb{F}[f](\omega)e^{i\omega x}d\omega$$

2.3 Properties and Formulas of Fourier Transform

a. **Linearity**: $\mathbb{F}[af+bg](\omega) = a\mathbb{F}[f](\omega)+b\mathbb{F}[g](\omega)$

b. If $f(x)\in\mathbb{R}$, then $F(-\omega) = \overline{F(\omega)}$

c. **Displacement formula**: if $g(x) = f(ax+b)$ then:

1. $\mathbb{F}[g](\omega) = \dfrac{1}{|a|}e^{\frac{i\omega b}{a}}\mathbb{F}[f]\left(\dfrac{\omega}{a}\right)$

2. $\mathbb{F}[e^{icx}f(x)](\omega) = \mathbb{F}[f](\omega-c)$

3. **Modulation formulas**:

$$\mathbb{F}[f(x)\cos cx](\omega) = \frac{\mathbb{F}[f](\omega-c)+\mathbb{F}[f](\omega+c)}{2}$$

$$\mathbb{F}[f(x)\sin cx](\omega) = \frac{\mathbb{F}[f](\omega-c)-\mathbb{F}[f](\omega+c)}{2i}$$

d. **Derivative formula**: If function $f(x)$ is continuous, $f,f'\in G(\mathbb{R})$ and $\lim_{x\to\pm\infty}f(x) = 0$, then

$$\mathbb{F}[f'](\omega) = i\omega\mathbb{F}[f](\omega) \ , \ \mathbb{F}[f^{(n)}](\omega) = (i\omega)^n\mathbb{F}[f](\omega)$$

e. If $f\in G(\mathbb{R})$ and $\int_{-\infty}^{\infty}|xf(x)|dx < \infty$, then Fourier transform of f is continuously differentiable and there holds

$$\mathbb{F}[xf(x)](\omega) = i\frac{d}{d\omega}\mathbb{F}[f](\omega)$$

$$\mathbb{F}[x^n f(x)](\omega) = i^n\frac{d^n}{d\omega^n}\mathbb{F}[f](\omega)$$

f. **Plancherel formula**: If $f \in G(\mathbb{R})$ and $\int\limits_{-\infty}^{\infty} |f(x)|^2\, dx < \infty$, then

$$\int\limits_{-\infty}^{\infty} |\mathbb{F}[f](\omega)|^2\, d\omega < \infty$$

and there holds $\int\limits_{-\infty}^{\infty} |f(x)|^2\, dx = \int\limits_{-\infty}^{\infty} |\mathbb{F}[f](\omega)|^2\, d\omega$

g. **Generalized Plancherel Formula**:

If $f, g \in G(\mathbb{R})$, $\int\limits_{-\infty}^{\infty} |f(x)|^2 < \infty$ and $\int\limits_{-\infty}^{\infty} |g(x)|^2 < \infty$, then

$$\int\limits_{-\infty}^{\infty} f(x)\overline{g(x)}\, dx = \int\limits_{-\infty}^{\infty} \mathbb{F}[f](\omega)\,\overline{\mathbb{F}[g](\omega)}\, d\omega$$

h. **A convolution** of two functions f, g is

$$(f*g)(x) = \int\limits_{-\infty}^{\infty} f(x-y)g(y)\, dy = \int\limits_{-\infty}^{\infty} f(y)g(x-y)\, dy = (g*f)(x)$$

i. $\mathbb{F}[f*g](\omega) = \sqrt{2\pi}\, \mathbb{F}[f](\omega) \cdot \mathbb{F}[g](\omega)$

2.4   Fourier Transforms Table

$f(t) = \dfrac{1}{\sqrt{2\pi}} \displaystyle\int_{-\infty}^{\infty} \mathbb{F}(\omega)\, e^{i\omega t}\, d\omega$	$F(\omega) = \dfrac{1}{\sqrt{2\pi}} \displaystyle\int_{-\infty}^{\infty} f(t)\, e^{-\omega t}\, dt$
$f(t)$	$F(\omega)$
$\dfrac{\sin \alpha t}{t}$, $\alpha > 0$	$\sqrt{\dfrac{\pi}{2}}$, $\lvert \omega \rvert < \alpha$ 0, $\lvert \omega \rvert > \alpha$
$\begin{cases} e^{i\alpha t}, & p < t < q \\ 0, & t < p,\ t > q \end{cases}$	$\dfrac{i}{\sqrt{2\pi}} \cdot \dfrac{e^{iq(\alpha-\omega)} - e^{ip(\alpha-\omega)}}{\omega - \alpha}$, $\omega \neq \alpha$ $\dfrac{q-p}{\sqrt{2\pi}}$, $\omega = \alpha$
$\begin{cases} e^{-ct+i\alpha t}, & t > 0,\ c > 0 \\ 0, & t < 0 \end{cases}$	$\dfrac{i}{\sqrt{2\pi}\,(\alpha - \omega + ic)}$
$e^{-\frac{t^2}{2}}$	$e^{-\frac{\omega^2}{2}}$
e^{-pt^2} , $\operatorname{Re} p > 0$	$\dfrac{1}{\sqrt{2p}}\, e^{-\frac{\omega^2}{4p}}$
$e^{i\alpha t^2}$, $\alpha > 0$	$\dfrac{1}{\sqrt{2\alpha}}\, e^{-i\left(\frac{\omega^2}{4\alpha} - \frac{\pi}{4}\right)}$
$\cos \alpha t^2$, $\alpha > 0$	$\dfrac{1}{\sqrt{2\alpha}} \cos\left(\dfrac{\omega^2}{4\alpha} - \dfrac{\pi}{4}\right)$
$\sin \alpha t^2$, $\alpha > 0$	$\dfrac{1}{\sqrt{2\alpha}} \sin\left(\dfrac{\pi}{4} - \dfrac{\omega^2}{4\alpha}\right)$
$\lvert t \rvert^{-s}$, $0 < \operatorname{Re} s < 1$	$\sqrt{\dfrac{2}{\pi}}\, \Gamma(1-s) \sin\dfrac{s\pi}{2} \cdot \lvert \omega \rvert^{s-1}$
$\dfrac{1}{\sqrt{\lvert t \rvert}}$	$\dfrac{1}{\sqrt{\lvert \omega \rvert}}$

Chapter 3
Laplace Transformation Formulas

3.1   Definition

a. Let $f(x)$ be a piecewise continuous function $f(x)$: $[0,\infty) \to \mathbb{C}$. Function $L[f](s) = \int_0^\infty e^{-st}f(t)\,dt$ of real variable s is called **Laplace transform** of $f(x)$.

b. If there exist real constants α and K such that $|f(t)| \le Ke^{\alpha t}$, $t \ge 0$, then $L[f](s)$ is defined for every $s > \alpha$.

3.2   Formulas of Laplace Transform

a. **Linearity**
$$L[\alpha f(s) + \beta g(s)] = \alpha L[f](s) + \rho L[f]s$$

b. **Differential formula**:
$$L[f'](s) = sL[f] - f(0)$$

c. **n-th order differential:**
$$L[f^{(n)}](s) = s^n L[f](s) - s^{n-1}f(0) - s^{n-2}f'(0) - \ldots - f^{(n-1)}(0)$$

d. $L[t^n f(t)](s) = (-1)^n \dfrac{d^n}{ds^n} L[f](s)$

e. $L[e^{at}f(t)](s) = L[f](s-a)$, $a \in \mathbb{R}$

f. $L[f(at)](s) = \dfrac{1}{a}L[f]\left(\dfrac{s}{a}\right)$, $a > 0$

g. If $f(t), t \geq 0$ is a periodic function with period T, then

$$L[f](s) = \frac{1}{1-e^{-Ts}} \int_0^T e^{-st} f(t) dt$$

3.3   Heaviside Step Function

For all real positive c, the function $u_c(t) = \begin{cases} 1, c \leq t \\ 0, 0 \leq t < c \end{cases}$ is

Heaviside step function.

a. $L[u_c](s) = \frac{1}{s} e^{-cs}$, $s > 0$

b. If $L[f](s)$ is Laplace transform of $f(x)$ and

$u_c(t) f(t-c) = \begin{cases} 0, \ 0 \leq t < c \\ f(t-c), t \geq c \end{cases}$, then

$L[u_c(t) f(t-c)](s) = e^{-cs} L[f](s)$.

c. If $f(t) = \begin{cases} 1 \ , \ t \in [a,b] \\ 0 \ , \ t \notin [a,b] \end{cases}$, $a > 0$, then

$L[f](s) = \frac{1}{s}(e^{-as} - e^{-bs})$.

3.4   Dirac Delta Function

a. Let a be a given real number. "Function" δ_a, holding $\int_A f(t) \delta_a(t) dt = f(a)$ for every function f continuous in the neighborhood of a, and for all set A containing a neighborhood of a, is called the **Dirac delta function** on a.

b. δ_a only exists as a function of functions. One common description of δ_a is a "limit of a process". For every $c > 0$, let us define function $\sigma_c(t) = \begin{cases} \dfrac{1}{2c}, \ a-c < t < a+c \\ 0, \ t \notin (a-c, a+c) \end{cases}$.

Properties of $\sigma_c(t)$:

1. $\int\limits_{-\infty}^{\infty} \sigma_c(t)\,dt = 1$

2. $\sigma_c(t) \geq 0$, for every $t \in \mathbb{R}$

3. $\lim\limits_{c \to 0^+} \sigma_c(t) = 0$, for every $t \neq a$

Function δ_a can be perceived as equal to the so-called limit $\lim\limits_{c \to 0^+} \sigma_c(t)$.

c. $\int\limits_{-\infty}^{\infty} f(t)\delta_a(t)\,dt = f(a)$

d. $L[\delta_a](s) = e^{-as}$, $a > 0$

3.5   Convolution (see 2.3 h,j)

a. $(f*g)(t) = \int\limits_0^t f(t-y)g(y)\,dy$

b. $(f*g)(t) = (g*f)(t)$

c. If there exist constants a, K_1, and K_2 such that $|f(t)| \leq K_1 e^{at}$ and $|g(t)| \leq K_2 e^{at}$ for every $t \geq 0$, then $|(f*g)(t)| \leq K_1 K_2 t e^{at}$, $t \geq 0$, and there holds

$$L[f*g](s) = L[f](s) \cdot L[g](s) \ , \ s > a_.$$

3.6   Table of Laplace Transforms

a, α are real numbers and $s > 0$

	$F(s) = L\{f(t)\}$	$f(t)$, $t > 0$
1	$\dfrac{1}{s}$	1
2	$\dfrac{1}{s^n}$, $n = 1, 2, \ldots$	$\dfrac{t^{n-1}}{(n-1)!}$
3	$\dfrac{1}{\sqrt{s}}$	$\dfrac{1}{\sqrt{\pi t}}$
4	$\dfrac{1}{s^{3/2}}$	$2\sqrt{\dfrac{t}{\pi}}$
5	$\dfrac{1}{s^\alpha}$, $\alpha > 0$	$\dfrac{t^{\alpha-1}}{\Gamma(\alpha)}$ (See V.3.3)
6	$\dfrac{1}{s-z}$, $s > \operatorname{Re} z$	e^{zt} , $z \in \mathbb{C}$
7	$\dfrac{1}{s-a}$, $s > a$	e^{at}
8	$\dfrac{1}{(s-a)^2}$	te^{at}
9	$\dfrac{1}{(s-a)^n}$, $n = 1, 2, \ldots$	$\dfrac{1}{(n-1)!} t^{n-1} e^{at}$
10	$\dfrac{1}{(s-a)^\alpha}$, $\alpha > 0$	$\dfrac{1}{\Gamma(\alpha)} t^{\alpha-1} e^{at}$
11	$\dfrac{1}{s^2 + a^2}$	$\dfrac{1}{a} \sin at$
12	$\dfrac{s}{s^2 + a^2}$	$\cos at$

	$F(s) = L\{f(t)\}$	$f(t)$, $t > 0$	
13	$\dfrac{1}{s(s^2 + a^2)}$	$\dfrac{1}{a^2}(1 - \cos at)$	
14	$\dfrac{1}{s^2(s^2 + a^2)}$	$\dfrac{1}{a^3}(at - \sin at)$	
15	$\dfrac{1}{(s^2 + a^2)^2}$	$\dfrac{1}{2a^3}(\sin at - at \cos at)$	
16	$\dfrac{s}{(s^2 + a^2)^2}$	$\dfrac{t}{2a}\sin at$	
17	$\dfrac{s^2}{(s^2 + a^2)^2}$	$\dfrac{1}{2a}(\sin at + at \cos at)$	
18	$\dfrac{1}{s^4 + 4a^4}$	$\dfrac{1}{4a^3}(\sin at \cdot \cosh at - \cos at \cdot \sin h\, at)$	
19	$\dfrac{s}{s^4 + 4a^4}$	$\dfrac{1}{2a^2}\sin at \cdot \sin h\, at$	
20	$\dfrac{1}{\sqrt{s^2 + a^2}}$	$J_0(at)$	(see XI.3.5)
21	$\dfrac{s}{(s - a)^{3/2}}$	$\dfrac{1}{\sqrt{\pi t}}e^{at}(1 + 2at)$	
22	$\dfrac{1}{\sqrt{s}}e^{-\alpha/s}$	$\dfrac{1}{\sqrt{\pi t}}\cos 2\sqrt{\alpha t}$	
23	$\dfrac{1}{s^{3/2}}e^{\alpha/s}$	$\dfrac{1}{\sqrt{\pi \alpha}}\sinh 2\sqrt{\alpha t}$	
24	$\dfrac{1}{s}\ln s$	$-\ln t - \gamma$, $(\gamma \approx 0.5772)$	

	$F(s) = L\{f(t)\}$	$f(t)$, $t > 0$
25	$\ln \dfrac{s-a}{s-b}$	$\dfrac{1}{t}(e^{bt} - e^{at})$
26	$\arctan \dfrac{a}{s}$	$\dfrac{1}{t}\sin at$

XIV. Partial Differential Equations (PDE)

Chapter 1
Introduction

a. A PDE is an equation relating an unknown multi-variable function and its partial differentials.

b. If the highest order of differentiation of a PDE is n, it is an n-**th order PDE.**

c. The general form of a PDE with 2 variables, x, y is

   1. First order: $F\left(x, y, u(x, y), u_x, u_y\right) = 0$

   2. Second order: $F\left(x, y, u(x, y), u_x, u_y, u_{xx}, u_{xy}, u_{yy}\right) = 0$

   when $u(x, y)$ is an unknown function and F is a given function.

d. Function $u(x, y)$ is a **solution** or an **integral of** n-**th order PDE** in domain D if u is n times partially differentiable in D, and, substituting it in the equation, one attain an identity in D.

   Example: The solution of $xu_x + yu_y = u - \dfrac{9}{u}$ is:

$u(x,y) = \sqrt{9-x^2-y^2}$ in domain $x^2+y^2 < 9$

$u(x,y) = \sqrt{x^2+9}$ in domain $\mathbb{R}^2$

$u(x,y) = \sqrt{xy+9}$ in domain $xy > -9$.

e. A PDE solution is called an **integral surface.**

f. A **general solution** of a PDE includes all solutions of the equation.

Example: A general solution of $u_{xy} = \sin x + y$ is

$u(x,y) = -y\cos x + \tfrac{1}{2}xy^2 + \varphi_1(x) + \varphi_2(y)$

where $\varphi_1(x)$ and $\varphi_2(y)$ are arbitrary differentiable functions.

g. $F[u] = F\big(x_1,...,x_k, u(x_1,...x_k), u_{x_1},..., u_{x_k},...\big) = f(x_1,...,x_k)$
   is a linear PDE **if** $F[\alpha u] = \alpha F[u]$ **and**
   $F[u+v] = F[u] + F[v]$ **(see 4.1).**

Chapter 2
First-Order Linear PDE

$$a(x,y)u_x + b(x,y)u_y + c(x,y)u = f(x,y) \qquad (*)$$

when $u(x,y)$ is an unknown function piecewise differentiable by x and y, functions $a(x,y)$, $b(x,y)$, $c(x,y)$, $f(x,y)$ are continuous in D and $a(x,y)$, $b(x,y)$ are not vanished together.

2.1   General Solution

Finding the solutions systematically:

a. Construct a **characteristic equation**

$$\frac{dy}{dx} = \frac{b(x,y)}{a(x,y)} \quad , \quad a(x,y) \neq 0$$

b. The characteristic equation's solutions $t(x,y) = c_0$ are called **characteristic curves** of (*)

c. Choose arbitrary function $s(x,y)$ of class C^1 such that holds

$$\begin{vmatrix} s_x & s_y \\ t_x & t_y \end{vmatrix} \neq 0 \ , \ (x,y) \in D$$

d. Changing variables $t = t(x,y)$, $s = s(x,y)$ in (*), we get a partially differential equation with respect to w: $(as_x + bs_y)w_s + cw = f$ whose solution is $w = w(s,t)$.

e. General solution of (*) is $u(x,y) = w\big(s(x,y), t(x,y)\big)$.

2.2   Cauchy Problem

a. Cauchy problem for a PDE consists of equation (*) and initial condition

$$x = x(t), \ y = y(t), \ u\big(x(t), \ y(t)\big) = \varphi(t), \ \alpha \le t \le \beta \quad (**)$$

b. If initial condition (**) is such that for every $\alpha < t < \beta$ vectors $\big(x(t), y(t)\big)$ and $\big(a(x(t),y(t)), \ b(x(t),y(t)\big)$ are not parallel, then Cauchy problem has a unique integral surface (see 3.2.b).

Chapter 3
Quasi-Linear PDE

$$a(xy,u)u_x + b(x,y,u)u_y = c(x,y,u) \quad (*)$$

when $a(x,y,u)$, $b(x,y,u)$, $c(x,y,u)$ are of class C^1 (see VII.4.3).

Solution $u = u(x,y)$ or $F(x,y,u) = C$ written in its implicit form is called an **integral surface**.

3.1   General Solution

a. Surface $S = \{(x,y,u) : F(x,y,u) = C\}$ is an integral surface of (*) if, and only if, in every point $M(x,y,u)$ on S, there holds

$$\nabla F \cdot \big(a(M), b(M), c(M)\big) = 0$$

b. An integral surface consists of characteristic curves

$$\sigma(t) = \{x(t), y(t), u(t)\} , \ \alpha < t < \beta$$

Therefore, vectors $\big(a(M), b(M), c(M)\big)$ and $\big(x'(t), y'(t), u'(t)\big)$ are parallel.

c. To find out all characteristic curves, we solve characteristic equation system

$$\frac{dx}{a(x,y,u)} = \frac{dy}{b(x,y,u)} = \frac{dz}{c(x,y,u)} \quad or \quad \begin{cases} \dfrac{dx}{dt} = a(x,y,u) \\[2mm] \dfrac{dy}{dt} = b(x,y,u) \quad (**) \\[2mm] \dfrac{dz}{dt} = c(x,y,u) \end{cases}$$

d. **Theorem**: If a general solution of (**) is given as the intersection of the two surfaces $F_1(x,y,u) = C_1$ and $F_2(x,y,u) = C_2$, then for every continuous function $\Phi(t,w)$ with partial continuous derivatives, the surface

$$\Phi(C_1,C_2) = \Phi\big(F_1(x,y,u),F_2(x,y,u)\big) = 0$$

is an integral surface of (*), for every arbitrary choice of parameters C_2,C_1.

3.2 Cauchy Problem

a. Finding integral surface $S:\{x(t,s),y(t,s),u(t,s)\}$ of PDE (*), passing through characteristic curve $\sigma_0(s) = \{x(0,s),y(0,s),u(0,s)\}$, $\alpha < s < \beta$ is a **Cauchy problem.**

b. **Existence and Uniqueness Theorem**: If, in PDE (*)

1. $a(x,y,u)$, $b(x,y,u)$, $c(x,y,u) \in C^1$ in solid V .

2. $\sigma_0(s) = \{x(0,s),y(0,s),u(0,s)\}$, $\alpha < s < \beta$ is a smooth, simple curve (see XII. 1.4).

3. There holds the **transversality** criterion

$$\begin{vmatrix} x'(0,s) & y'(0,s) \\ a\big(x(0,s),y(0,s),u(0,s)\big) & b\big(x(0,s),y(0,s),u(0,s)\big) \end{vmatrix} \neq 0 \;,\; \alpha < s < \beta$$

That is, if vectors $(x'(0,s), y'(0,s))$ and (a,b) are not parallel, then there exists a unique integral surface S_0 containing $\sigma_0(s)$.

c. If the transversality criterion of b. does not hold, then, when the rank of matrix

$$A = \begin{pmatrix} x'(0,s) & y'(0,s) & u'(0,s) \\ a\big(x(0,s), y(0,s), u(0,s)\big) & b\big(x(0,s), y(0,s), u(0,s)\big) & c\big(x(0,s), y(0,s), u(0,s)\big) \end{pmatrix}$$

is 1, the Cauchy problem has an infinite number of solutions. When $\operatorname{rank} A = 2$, the Cauchy problem has no solutions.

d. **Lagrange's Method of Solution**:

If the general solution of characteristic equation (**) is the line of intersection between surfaces $F_1(x,y,u) = C_1$ and $F_2(x,y,u) = C_2$, and initial condition $\sigma_0(s)$ is also given as the intersection of surfaces $\Phi_1(x,y,u) = 0$, $\Phi_2(x,y,u) = 0$, then, out of these 4 equations, we extract u, y, x and get a relation between C_1 and C_2, which is $\Phi(C_1, C_2)$.

Substituting C_1 and C_2, F_1 and F_2. We get the required solution $\Phi\big(F_1(x,y,u), F_2(x,y,u)\big) = 0$ (see 3.1.d).

Chapter 4
Second-Order Linear PDE in Two Variables

4.1   Basic Concepts

a. The equation

$$L[u] = a(x,y)u_{xx} + 2b(x,y)u_{xy} + c(x,y)u_{yy} + \qquad (*)$$

$$+d(x,y)u_x + e(x,y)u_y + f(x,y)u = g(x,y)$$

when a,b,c,d,e,f,g are given functions and $u = u(x,y)$ is the unknown function, is the second-order PDE.

b. For every two functions u_1 and u_2 from class C^2, and for every two constants α, β there holds

$$L[\alpha u_1 + \beta u_2] = \alpha L[u_1] + \beta L[u_2]$$

c. A linear combination of n functions $u_1, \ldots u_n$ for which exist $L[u_i]$, holds

$$L[\sum_{i=1}^{n} \alpha_i u_i] = \sum_{i=1}^{n} \alpha_i L[u_i]$$

d. **The Superposition Principle**:

1. If $u_1, u_2, \ldots, u_n$ are n solutions of the equation $L[u] = 0$, then every linear combination of $\sum_{i=1}^{n} \alpha_i u_i$ is also a solution of that equation.

2. Let $\{u_i\}_{i=1}^{\infty}$ be an infinite sequence of solutions of $L[u]=0$. If the series $\sum\limits_{i=1}^{\infty} \alpha_i u_i$:

I. Uniformly converges to function u.

II. Twice differentiable and the derivative of the sum is equal to the sum of derivatives, (see VI.4.4), then also u is a solution of $L[u]=0$ and

$$L[u] = L[\sum_{i=1}^{\infty} \alpha_i u_i] = \sum_{i=1}^{\infty} \alpha_i L[u_i] = 0$$

4.2   Classification

a. The expression $a(x,y)u_{xx} + 2b(x,y)u_{xy} + c(x,y)u_{yy}$ is called the **principal part** of (*)

b. The equations

$$\frac{dy}{dx} = \frac{b+\sqrt{b^2-ac}}{a} \quad , \quad \frac{dy}{dx} = \frac{b-\sqrt{b^2-ac}}{a} \qquad (**)$$

are **characteristic equations**, the solutions of which, $\varphi(x,y) = C_1$ and $\psi(x,y) = C_2$ are **characteristic curves**.

c. If $b^2 - ac > 0$ on D, then (*) is said to be a **hyperbolic equation** on D.

d. If $b^2 - ac = 0$ on D, then (*) is said to be a **parabolic equation** on D.

e. If $b^2 - ac < 0$ on D, then (*) is said to be an **elliptic equation** on D.

4.3   Reducing PDE into a Canonical Form

a. Change of Variables

Let $\eta = \psi(x,y), \xi = \varphi(x,y)$ be continuous functions, with continuous first-order partial derivatives and the Jacobian

$$\begin{vmatrix} \varphi_x & \varphi_y \\ \psi_x & \psi_y \end{vmatrix} \neq 0 \text{ in domain } D.$$

Changing variables, we denote:

$$w(\xi,\eta) = u\big(x(\xi,\eta), y(\xi,\eta)\big)$$

Using the Chain rule, we get

$$u_x = w_\xi \xi_x + w_\eta \eta_x$$

$$u_y = w_\xi \xi_y + w_\eta \eta_y$$

$$u_{xx} = w_{\xi\xi}\xi_x^2 + 2w_{\xi\eta}\xi_x\eta_x + w_{\eta\eta}\eta_x^2 + w_\xi \xi_{xx} + w_\eta \eta_{xx}$$

$$u_{xy} = w_{\xi\xi}\xi_x\xi_y + w_{\xi\eta}(\xi_x\eta_y + \xi_y\eta_x) + w_{\eta\eta}\eta_x\eta_y + w_\xi \xi_{xy} + w_\eta \eta_{xy}$$

$$u_{yy} = w_{\xi\xi}\xi_y^2 + 2w_{\xi\eta}\xi_y\eta_y + w_{\eta\eta}\eta_y^2 + w_\xi \xi_{yy} + w_\eta \eta_{yy}$$

b. If $b^2 - ac > 0$, the characteristic equations (**) have two sets of characteristic curves $\varphi(x,y) = C_1$, $\psi(x,y) = C_2$.
By change of variables $\xi = \varphi(x,y)$, $\eta = \psi(x,y)$, and substitution in (*), we get $w_{\xi\eta} + F(\xi,\eta,w,w_\xi,w_\eta) = 0$, a **canonical form of a hyperbolic equation**.

c. If $b^2 - ac = 0$ in D, then equation (*) has one characteristic curve, $\varphi(x,y) = C$. We choose new variables $\xi = \varphi(x,y)$ and η such that the Jacobian will be nonzero in every point of D. Substituting in (*), we get $w_{\xi\xi} + F(\xi,\eta,w,w_\xi,w_\eta) = 0$, a **canonical form of a parabolic equation**.

d. If $b^2 - ac < 0$ in D, then the characteristic equations have two complex solutions. If $\varphi(x,y) + i\psi(x,y) = C$ is a solution of (**), we choose new variables $\xi = \varphi(x,y)$,

$\eta = \psi(x, y)$.   Substituting   in   (*),   we   get $u_{\xi\xi} + u_{\eta\eta} + F(\xi, \eta, w, w_\xi, w_\eta) = 0$, a **canonical form of an elliptic equation**.

Chapter 5
One-Dimensional Wave Equation

$$u_{tt} - c^2 u_{xx} = F(x, t) \ , \ -\infty \le a < x < b \le \infty \ , \ t > 0 \quad (*)$$

when $c \in \mathbb{R}$ is the **velocity of wave propagation.**

If $F(x, t) \equiv 0$, then it is a **homogeneous equations**.

5.1   General Solution of Homogeneous Equation
$u_{tt} - c^2 u_{xx} = 0$

a. $x - ct = \text{const}$, $x + ct = \text{const}$ are **characteristic lines**.

b. By change of variables $\xi = x + ct$, $\eta = x - ct$ we get the canonical form of a wave equation $-4c^2 w_{\xi\eta} = 0$. Its solution is $w(\xi, \eta) = F(\xi) + G(\eta)$, when functions F, G are arbitrary, continuous, and have partial derivatives continuous up to second order.

c. $u(x, y) = F(x + ct) + G(x - ct)$ is a general solution of (*).

d. $G(x - ct)$ describes a (rightwards) **forward wave** at speed c .
$F(x + ct)$ describes a (leftwards) **backward wave** at speed c .
General solution $u(x, y)$ is a superposition of an forward wave and a backward wave with speed c .

5.2 Vibrations of an Infinite String. D'Alembert's Formula

a. Equations system

$$u_{tt} - c^2 u_{xx} = 0 \ , \ -\infty < x < \infty \ , \ t > 0 \qquad (*)$$

$$u(x,0) = f(x) \ , \ u_t(x,0) = g(x) \ , \ -\infty < x < \infty \qquad (**)$$

describes the amplitude of an (ideal) elastic infinite string vibration. Initial conditions $f(x)$ and $g(x)$ are given functions describing amplitude u and vibration velocity u_t at time $t = 0$.

b. $(*),(**)$ is called a **Cauchy problem**.

c. If $f(x) \in C^2$ and $g(x) \in C^1$, then Cauchy problem $(*),(**)$ has unique solution $u(x,t)$ given by **D'Alembert's formula**

$$u(x,t) = \frac{f(x-ct) + f(x+ct)}{2} + \frac{1}{2c} \int\limits_{x-ct}^{x+ct} g(s)\,ds \qquad (***)$$

d. If $f(x)$ is continuous, $f'(x)$, $g(x)$ are piecewise continuous, and $f(x) \notin C^2$ and/or $g(x) \notin C^1$, then there are points where the first and second order derivatives of $u(x,t)$ do not necessarily exist, and therefore, function $u(x,t)$ given by $(**)$ is not a solution of problem/ $(*)$, $(**)$.

In any point except these ones, $u(x,t)$ is a solution of the wave problem. In this case, we construct a **generalized solution** of the wave problem, the following way:

Choose two sequences $\{f_n(x)\}_{n=1}^{\infty}$, $\{g_n(x)\}_{n=1}^{\infty}$, when $f_n(t) \in C^2$, $g_n(x) \in C^1$, uniformly converging to $f(x)$ and $g(x)$, respectively, in $|x| < \infty$.

Let the sequence $\{u_n(x,t)\}_{n=1}^\infty$ be a solution of (*) holding.

$$\frac{\partial u_n(x,0)}{\partial t} = g_n(x) \ , \ u_n(x,0) = f_n(x)$$

Then, the sequence $\{u_n(x,t)\}_{n=1}^\infty$ uniformly converges to the generalized solution, $u(x,t)$.

5.3   Non-Homogeneous Wave Equation

a. The equations system

$$u_{tt} - c^2 u_{xx} = F(x,t) \ , \ -\infty < x < \infty, t > 0 \quad (*)$$

$$u(x,0) = f(x), u_t(x,0) = g(x) \ , \ -\infty < x < \infty \quad (**)$$

describes a vibration of an infinite string constrained by an external force $F(x,t)$. String amplitude $u(x,t)$ is dependent of initial conditions $f(x)$ and initial velocity $g(x)$ at time $t = 0$.

b. The characteristic triangle Δ

Through point (x_0, t_0), we draw 2 characteristic lines:

$$x - ct = x_0 - ct_0, \ x + ct = x_0 + ct_0$$

forming, together with the x-axis, a characteristic triangle (see illustration).

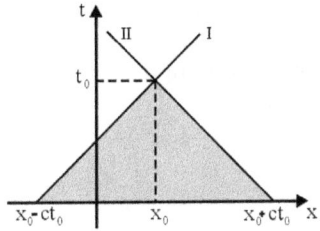

c. If, in (**), $f(x) = 0$ and $g(x) = 0$, then, the solution of (*) is

$$u_0(x,t) = \frac{1}{2c} \iint_\Delta F(s,\tau)\,ds\,d\tau = \frac{1}{2c} \int_0^t d\tau \int_{x-c(t-\tau)}^{x+c(t-\tau)} F(s,\tau)\,ds$$

d. If $f(x) \in C^2$, $g(x) \in C^1$ and functions $F(x,t)$ and $F_x(x,t)$ are continuous on $|x| < \infty$, $t \geq 0$, then the unique solution of Cauchy problem (*),(**) is

$$u(x,t) = \frac{f(x-ct)+f(x+ct)}{2} + \frac{1}{2c} \int_{x-ct}^{x+ct} g(s)\,ds + \frac{1}{2c} \iint_\Delta F(s,\tau)\,ds\,d\tau$$

5.4   Vibrations of Semi-Infinite String

a. Homogeneous cauchy problem:

$$u_{tt} - c^2 u_{xx} = 0 \ , \ x > 0, t > 0 \quad (*)$$

Initial condition:

$$u(x,0) = f(x) \ , \ u_t(x,0) = g(x) \ , \ x > 0 \quad (**)$$

Boundary condition:

$$u(0,t) = 0 \ , \ t > 0 \quad (***)$$

The solution is

$$u(x,t) = \begin{cases} \dfrac{f(x+ct)+f(x-ct)}{2} + \dfrac{1}{2c}\displaystyle\int_{x-ct}^{x+ct} g(s)\,ds \ , \ 0 < t < \dfrac{x}{c} \\[3mm] \dfrac{f(x+ct)-f(x-ct)}{2} + \dfrac{1}{2c}\displaystyle\int_{ct-x}^{x+ct} g(s)\,ds \ , \ t > \dfrac{x}{c} \end{cases}$$

b. The solutions of non-homogeneous problem

$$w_{tt} - c^2 w_{xx} = F(x,t) \ , \ x > 0, t > 0$$

with the same **initial condition (**)** or **boundary condition (***)** are $w(x,t) = u(x,t) + u_0(x,t)$

when $u(x,t)$ is the solution of homogeneous problem (see a.) and

$$u_0(x,t) = \begin{cases} \dfrac{1}{2c}\displaystyle\int_0^t d\tau \displaystyle\int_{x-c(t-\tau)}^{x+c(t-\tau)} F(s,\tau)\,ds & , \ 0 < t < \dfrac{x}{c} \\[3ex] \dfrac{1}{2c}\displaystyle\int_0^{t-\frac{x}{c}} d\tau \displaystyle\int_{-x+c(t-\tau)}^{x+a(t-\tau)} F(s,\tau)\,ds + \dfrac{1}{2c}\displaystyle\int_{t-\frac{x}{c}}^{t} d\tau \displaystyle\int_{x-c(t-\tau)}^{x+c(t-\tau)} F(s,\tau)\,ds & , \ t > \dfrac{x}{c} \end{cases}$$

c. If, in a Cauchy problem, the boundary condition is non-homogeneous, that is, (***) is replaced by $u_x(0,t) = H(t)$, $t > 0$, then, by substituting $u = w + xH(t)$, we get the non-homogeneous problem

$$w_{tt} - c^2 w_{xx} = xH''(t)$$

$$w(x,0) = f(x) - xH(0) \ , \ w_t(x,0) = g(x) + xH'(0) \ , \ w_x(0,t) = 0$$

5.5   Vibrationof a Finite String (Both Ends Fixed)

The finite string wave equation is

$$u_{tt} - c^2 u_{xx} = \Phi(x,t) \ , \ 0 < x < \ell, t > 0$$

Initial condition is: $u(x,0) = f(x)$, $u_t(x,0) = g(x)$, $0 \le x \le \ell$

Boundary condition is: $u(0,t) = 0$, $u(\ell,t) = 0$, $t \ge 0$

To use D'Alembert's formula (see 5.3), we extend functions $f(x)$, $g(x)$, $\Phi(x,t)$ on all x-axis, through interval ends $x = 0$ and $x = \ell$, to odd and periodic functions with a 2ℓ period, that is

$$F(x) = \begin{cases} f(x) \, , \, 0 < x < \ell \\ -f(-x) \, , \, -\ell < x < 0 \end{cases} , \quad F(x+2\ell) = F(x)$$

Similarly, we construct functions $G(x)$ and $\Phi(x,t)$.

Using the D'Alembert's formula mentioned in 5.3 to find the solution of the problem.

The reduction of this solution to $0 \le x \le \ell$, $t \ge 0$ is the solution of the given problem.

If, in addition, $f(x), g(x) \in C^2$, $0 \le x \le \ell$ satisfy the compatibility condition

$$f(0) = f(\ell) = 0 \, , \, f''(0) = f''(\ell) = 0 \, , \, g(0) = g(\ell) = 0$$

and $\Phi(x,t) \in C^1$, then Cauchy problem has a unique solution $u(x,y) \in C^2$.

Chapter 6
Method of Separation of Variables

6.1   Solution of the Wave Equation

$$u_{tt} - c^2 u_{xx} = 0 \, , \, 0 < x < \ell \, , \, t > 0 \quad (*)$$

with one of the following boundary value problems:

a. **Dirichlet problem** $u(0,t) = u(\ell,t) = 0$, $t \ge 0$

b. **Neumann problem** $u_x(0,t) = u_x(\ell,t) = 0$, $t \ge 0$

c. **Mixed problem** $u_x(0,t) = u(\ell,t) = 0$, $t \ge 0$

d. **Mixed problem** $u(0,t) = u_x(\ell,t) = 0$, $t \ge 0$;

and **initial condition**:
$$u(x,0) = f(x) \ , \ u_t(x,0) = g(x) \ , \ 0 \le x \le \ell$$

We look for a non-trivial solution of (*) in the form of $u(x,t) = X(x) \cdot T(t)$ holding one of the four boundary conditions. As a result of substitution in (*) and separation of variables by division by $X(x) \cdot T(t)$, we get

$$\frac{T''(t)}{c^2 T(t)} = \frac{X''(x)}{X(x)} = -\lambda$$

From that equation, follows the equations system

$$\begin{cases} X''(x) + \lambda X(x) = 0 \ , \ 0 < x < \ell \\ T''(t) + \lambda c^2 T(t) = 0 \ , \ t > 0 \end{cases}$$

For the solution $u(x,t) = X(x) \cdot T(t)$ to hold a boundary condition, for example, a Dirichlet condition, there must hold $X(0) = 0 \ , \ X(\ell) = 0$.

To find $X(x)$, we get Sturm-Liouville problem (see XI.5).

$$\begin{cases} X''(x) + \lambda X(x) = 0 \\ X(0) = 0 \ , \ X(\ell) = 0 \end{cases}$$

With eigenvalues $\lambda_n = \left(\dfrac{n\pi}{\ell} \right)^2$, $n = 1, 2, \ldots$ and

eigenfunctions $X_n(x) = \sin \dfrac{n\pi x}{\ell}$.

For every eigenvalue λ_n we get the equation

$$T''_n(t) + \frac{n^2 \pi^2 c^2}{\ell^2} T_n(t) = 0 \ , \ t > 0$$

Its solutions are $T_n(t) = A_n \cos \dfrac{n\pi c}{\ell} t + B_n \sin \dfrac{n\pi c}{\ell} t$

Therefore, we obtain particular solutions

$$u_n(x,t) = X_n(x)T_n(t) = \sin\frac{n\pi}{\ell}x\left(A_n\cos\frac{n\pi c}{\ell}t + B_n\sin\frac{n\pi c}{\ell}t\right),$$

$$n = 1,2,\ldots$$

Following the generalized superposition principle, function

$$u(x,t) = \sum_{n=1}^{\infty}\left(A_n\cos\frac{n\pi c}{\ell}t + B_n\sin\frac{n\pi c}{\ell}t\right)\sin\frac{n\pi}{\ell}x$$

is the generalized solution of the wave problem (see XIV.5.2) when coefficients A_n and B_n are deduced from the initial condition

$$u(x,0) = \sum_{n=1}^{\infty}A_n\sin\frac{n\pi}{\ell}x = f(x),$$

$$u_t(x,0) = \sum\frac{n\pi c}{\ell}B_n\sin\frac{n\pi}{\ell}x = g(x)$$

That is, A_n are Fourier coefficients of $f(x)$ and B_n are Fourier coefficients of the expansion of $g(x)$ to a Fourier series by cosines (see XIII, 1.5. 1.6).

6.2   Solution of Homogeneous Heat Equation

$$u_t - ku_{xx} = 0 , \quad 0 < x < \ell , \quad t > 0$$

Boundary condition: $u(0,t) = u(\ell,t) = 0$, $t \geq 0$
Initial condition: $u(x,0) = f(x)$, $0 \leq x \leq \ell$
Compatibility condition: $f(0) = f(\ell) = 0$
We look for a solution in the form of $u(x,t) = X(x)\cdot T(t)$.
Substituting, we get the Sturm-Liouville problem

$$X''(x) + \lambda X(x) = 0 , \quad 0 < x < \ell$$

$$x(0) = x(\ell) = 0$$

with eigenvalues $\lambda_n = \left(\dfrac{n\pi}{\ell}\right)^2$, $n = 1, 2, \ldots$ and eigenfunctions

$X_n(x) = \sin\dfrac{n\pi x}{\ell}$.

Function $T_n(t) = B_n e^{-k\left(\frac{n\pi}{\ell}\right)^2 t}$ is a solution of the equation $T''(t) + \lambda k T = 0$. Therefore, the solution of the heat equation is

$$u(x,t) = \sum_{n=1}^{\infty} B_n \sin\frac{n\pi}{\ell} x \cdot e^{-k\left(\frac{n\pi}{\ell}\right)^2 t}$$

From the initial condition, we get

$$u(x,0) = \sum_{n=1}^{\infty} B_n \sin\frac{n\pi x}{\ell} = f(x)$$

Therefore, $B_n = \dfrac{2}{\ell}\displaystyle\int_0^{\ell} f(x)\sin\dfrac{n\pi x}{\ell}\,dx$ are Fourier coefficients of function $f(x)$.

6.3   General Heat Equation and the Maximum Principle

a. **The Maximum Principle**:

If $u(x,t)$ is a continuous solution in rectangle $0 \le x \le \ell$, $0 \le t \le T$ of heat equation $u_t = ku_{xx}$, then $u(x,t)$ attains its maximum on the base $t = 0$ or sides $x = 0$ and $x = \ell$.

b. Heat problem

$$t > 0,\ 0 < x < \ell,\ u_t - ku_{xx} = F(x,t)$$

$$u(x,0) = f(x)\ ,\ 0 \le x \le \ell$$

$$u(0,t) = a(t)\ ,\ u(\ell,t) = b(t)\ ,\ t \ge 0$$

has a continuous solution on $0 \le x \le \ell$, and its is unique.

c. Heat problem

$$u_t - ku_{xx} = F(x,t) \ , \ -\infty < x < \infty \ , \ t > 0$$

$$u(x,0) = f(x) \ , \ -\infty < x < \infty$$

has a unique bounded and continuous solution.

If $F(x,t) \equiv 0$, the solution is

$$u(x,t) = \frac{1}{2\sqrt{\pi kt}} \int_{-\infty}^{\infty} \exp\left(\frac{-(x-\xi)^2}{4kt}\right) f(\xi) d\xi$$

Chapter 7
Laplace Equation

7.1   Introduction

Let D be a domain with boundary ∂D, $\mathbf{n}$ a normal to ∂D, and $\Delta = \dfrac{\partial^2}{\partial x^2} + \dfrac{\partial^2}{\partial y^2}$ a **Laplace operator**

a. Finding solution $u(x,y)$ in D of Laplace equation $\Delta u = u_{xx} + u_{yy} = 0$, $(x,y) \in D$.

with the boundary condition:

1. $u = f$, $(x,y) \in \partial D$ **Dirichlet problem**

2. $\dfrac{\partial u}{\partial \mathbf{n}} = f$, $(x,y) \in \partial D$ **Neumann problem**

3. $\dfrac{\partial u}{\partial \mathbf{n}} + hu = g$, $(x,y) \in \partial D$ **Robin problem**

b. The Dirichlet problem has a unique solution; the Neumann problem has a unique up to constant solution; the Robin problem has a unique solution when $h \geq 0$ and is not identically zero.

c. **Maximum Principle**: If $u(x,y)$ is a harmonic function (see XII.4.5), bounded on D, and continuous on $D \cup \partial D$ then $u(x,y)$ attains its maximum on ∂D.

7.2 Dirichlet Problem in a Rectangle

$$u_{xx} + u_{yy} = 0 , \quad 0 < x < a , \quad 0 < y < b$$

$$u(0,y) = f_1(y) , \quad u(a,y) = f_2(y) , \quad 0 \leq y \leq b$$

$$u(x,0) = f_3(x) , \quad u(x,b) = f_4(x) , \quad 0 \leq x \leq a$$

The solution is a superposition of 4 functions, u_1, u_2, u_3, u_4, each function u_i being a solution of a Dirichlet problem where boundary condition, except f_i, is identically zero.

Example: If $f_1 = f_2 = f_3 \equiv 0$, $f_4(x) \neq 0$ the problem is solved by separation of variables $u(x,y) = X(x) \cdot Y(y)$.

The generalized solution is $u(x,y) = \sum_{n=1}^{\infty} c_n \sin \dfrac{n\pi x}{a} \cdot \sinh \dfrac{n\pi y}{a}$

when $c_n \sinh \dfrac{n\pi b}{a}$ are Fourier coefficients of $f_4(x)$, and therefore

$$c_n = \left(\frac{2}{a} \int_0^a f_4(x) \sin \frac{n\pi x}{a} \, dx \right) \frac{1}{\sinh n\pi b/a}$$

7.3 Dirichlet Problem in a Disc

$$u_{xx} + u_{yy} = 0 , \quad x^2 + y^2 < a^2$$

$$u = f , \quad x^2 + y^2 = a^2$$

In polar coordinates $x = \rho\cos\theta$, $y = \rho\sin\theta$, $0 \le \theta \le 2\pi$, $0 < \rho \le a$ we get

$$u_{rr} + \frac{1}{r}u_r + \frac{1}{r^2}u_{\theta\theta} = 0 \text{ , } r < a$$

$$u(a,\theta) = f(\theta) \text{ , } 0 \le \theta \le 2\pi$$

This problem is solved by separation of variables

$$u(r,\theta) = R(r)\Theta(\theta)$$

when $u(r,\theta+2\pi) = u(r,\theta)$, which leads to $\Theta(-\pi) = \Theta(\pi)$, $\Theta'(-\pi) = \Theta'(\pi)$.

The final solution is

$$u(r,\theta) = \frac{a_0}{2} + \sum_{n=1}^{\infty} \frac{r^n}{a^n}(a_n \cos n\theta + b_n \sin n\theta) \quad (*)$$

when a_n and b_n are Fourier coefficients of $f(\theta)$ on $[-\pi,\pi]$

$$a_n = \frac{1}{\pi}\int_{-\pi}^{\pi} f(\theta)\cos n\theta d\theta \text{ , } n = 0,1,\ldots,$$

$$b_n = \frac{1}{\pi}\int_{-\pi}^{\pi} f(\theta)\sin n\theta d\theta \text{ , } n = 1,2,\ldots$$

(*) can be written in the form of **Poisson's integral formula**:

$$u(r,\theta) = \frac{1}{2\pi}\int_{-\pi}^{\pi} \frac{a^2 - r^2}{a^2 - 2ar\cos(\theta-\varphi) + r^2} f(\varphi)d\varphi .$$

7.4   Neumann Problem in a Disc

$$u_{rr} + \frac{1}{r}u_r + \frac{1}{r^2}u_{\theta\theta} = 0 \text{ , } r < a$$

$$\frac{\partial u}{\partial r}(a,\theta) = f(\theta) \text{ , } 0 \le \theta \le 2\pi, \text{ when } \int_0^{2\pi} f(\theta)d\theta = 0$$

The solution is $u(r,\theta) = \dfrac{\alpha_0}{2} + \dfrac{a}{\pi} \displaystyle\int_{-\pi}^{\pi} \left[\sum_{n=1}^{\infty} \dfrac{r^n}{a^n} \dfrac{\cos n(\theta-\varphi)}{n} \right] f(\varphi)\, d\varphi$

when α_0 is a constant.

XV. Combinatorics and Newton's Binomial

Chapter 1
Permutations

1.1   Permutations without Repetition

a. A **permutation** in n distinct objects (elements) is their possible arrangement in a line.

b. The difference between permutation in n distinct objects and another permutation of the same objects is in the order of its terms.

c. The total number of permutations of n objects is:

$P_n = P(n,n) = n!$, (n-factorial)

d. The number of **circular, or cyclic, permutations** is the number of all possible rearrangements of n distinct elements in a circle.

It is $P_n^* = (n-1)!$

1.2   Permutations with Repetition

a. A **permutation with repetition** is the linear arrangement of n objects, when n_1 objects are similar, another n_2 objects are similar, and the remaining $n_k,...,$ are similar, and when $n_1 + n_2 + ... + n_k = n$. It is denoted $P_n(n_1, n_2, ... n_k)$.

b. $P_n(n_1, n_2, ..., n_k) = \dfrac{n!}{n_1! \cdot n_2! \cdot ... \cdot n_k!}$

Chapter 2
Variations

2.1   Variations without Repetition

a. Any choice of k distinct objects out on n given distinct objects, when the order of choice is significant, is called a **variation**.

b. One variation of k distinct objects is different from another variation of k distinct objects in the order of terms of a least in one term.

c. The number of all different variations of k distinct objects out of n distinct objects is denoted as $P(n,k)$ and equals to

$$P(n,k) = n(n-1)(n-2)...(n-k+1) = \frac{n!}{(n-k)!} \quad (*)$$

d. **Multiplication principle:** In a given set of n distinct objects:

1. There are n possible choices for the first-place term.

2. For any of the first-place choices, there are just $n-1$ possible second-place choices, and therefore, there are $n(n-1)$ possible choices of two distinct objects.

3. For any of these $n(n-1)$ possible choices, there are $n-2$ possible third-place choices, and therefore, the number of possible different threesomes is $n(n-1)(n-2)$.

4. Following the same principle, we will find formula (*) of calculating the number of different variations of k out of n distinct objects.

Example: How many numbers of 5 different digits can you form using the digits $0,1,2,3,4,5,6,7,8,9$?

Solution: We arrange these 10 digits in 5 positions. The first on the left can be 9 digits, 0 excluded. The second on the left can also be 9 digits, including zero, but one digit had already been used. Third on the left can be 8 digits, fourth on the left, 7 digits, and the fifth on the left, 6 digits. Following the multiplication principle, the number of possible numbers is $9 \cdot 9 \cdot 8 \cdot 7 \cdot 6 = 27216$.

2.2   Variations with Repetition

a. A variation with repetition of k out of n distinct objects is a choice of k out on n given distinct objects, when the order of choice is significant, and any object can be chosen several times.

b. The number of variation with repetition of k out of n objects is denoted as V_n^k.

c. Using the multiplication principle, we get

$$V_n^k = \underbrace{n \cdot n \cdot \ldots \cdot n}_{k-\text{times}} = n^k$$

d. **Example**: How many numbers of 5 not necessarily different digits can you form using the digits $0,1,2,3,4$?

Solution: Using the multiplication principle, we get
$4 \cdot 5 \cdot 5 \cdot 5 \cdot 5 = 2500$

Chapter 3
Combination

a. The **combination** of k out of n distinct objects is a choice of k out of n given distinct objects, when the order of choice is insignificant and no object can be chosen more than one time.

b. One combination of k out of n distinct objects is different from another combination of k out of same n objects in at least one term.

c. The number of combinations is denoted as $C(n,k)$.

d. $C(n,k) = \dfrac{P(n,k)}{P(k,k)} = \dfrac{n!}{k!(n-k)!} = C_n^k = \dbinom{n}{k}$

Chapter 4
Newton's Binomial

a. Newton's Binomial formula:

$$(a+b)^n = C_n^0 a^n + C_n^1 a^{n-1} b + C_n^2 a^{n-2} b^2 + \ldots + C_n^n b^n = \sum_{k=0}^{n} C_n^k a^{n-k} b^k$$

$C_n^i = \dfrac{n!}{(n-i)!i!}$ are called **binomial coefficients**.

b. The term in the $k+1$ position is $T_{k+1} = C_n^k a^{n-k} b^k$.

c. $(a-b)^n = \sum_{k=0}^{n} (-1)^k C_n^k a^{n-k} b^k$

d. $C_n^0 + C_n^1 + C_n^2 + \ldots + C_n^n = 2^n$

e. $C_n^0 - C_n^1 + C_n^2 - C_n^3 + \ldots + (-1)^n C_n^n = 0$

f. **Pascal Triangle**, binomial coefficients arrangement:

$(a+b)^0$ | | | | | | | | 1 | | | | | | | |
$(a+b)^1$ | | | | | | | 1 | | 1 | | | | | |
$(a+b)^2$ | | | | | | 1 | | 2 | | 1 | | | | |
$(a+b)^3$ | | | | | 1 | | 3 | | 3 | | 1 | | | |
$(a+b)^4$ | | | | 1 | | 4 | | 6 | | 4 | | 1 | | |
$(a+b)^5$ | | | 1 | | 5 | | 10 | | 10 | | 5 | | 1 | |
$(a+b)^6$ | | 1 | | 6 | | 15 | | 20 | | 15 | | 6 | | 1 |
$(a+b)^7$ | 1 | | 7 | | 21 | | 35 | | 35 | | 21 | | 7 | | 1 |
$(a+b)^8$ | 1 | 8 | | 28 | | 56 | | 70 | | 56 | | 28 | | 8 | 1 |

A Pascal triangle is a triangular arrangement of numbers, the apex being 1, and the numbers on the sides are all 1. Every number of the triangle is a sum of two numbers above it.

XVI. Table of Integrals

Constant c is added to each integral

1.   Integrals of type $\int \dfrac{x^{\pm n}dx}{(a+bx)^m}$

1.1 $\int \dfrac{dx}{a+bx} = \dfrac{1}{b}\ln|a+bx|$

1.2 $\int \dfrac{dx}{(a+bx)^m} = \dfrac{-1}{(m-1)b(a+bx)^{m-1}}$, $m \geq 2$

1.3 $\int \dfrac{xdx}{a+bx} = \dfrac{1}{b}\left(x - \dfrac{a}{b}\ln|a+bx|\right)$

1.4 $\int \dfrac{xdx}{(a+bx)^2} = \dfrac{1}{b^2}\left(\dfrac{a}{a+bx} + \ln|a+bx|\right)$

1.5 $\int \dfrac{xdx}{(a+bx)^m} = \dfrac{1}{b^2}\left[\dfrac{-1}{(m-2)(a+bx)^{m-2}} + \dfrac{a}{(m-1)(a+bx)^{m-1}}\right]$, $m \geq 3$

1.6 $\int \dfrac{x^2dx}{a+bx} = \dfrac{1}{b}\left[\dfrac{x^2}{2} - \dfrac{a}{b}x + \left(\dfrac{a}{b}\right)^2 \ln|a+bx|\right]$

1.7 $\int \dfrac{x^2dx}{(a+bx)^2} = \dfrac{1}{b^2}\left[x - \dfrac{a}{b}\left(\dfrac{a}{a+bx} + 2\ln|a+bx|\right)\right]$

1.8 $\int \dfrac{x^2dx}{(a+bx)^3} = \dfrac{1}{b^3}\left[\dfrac{2a}{a+bx} - \dfrac{a^2}{2(a+bx)^2} + \ln|a+bx|\right]$

1.9 $\int \dfrac{x^n dx}{a+bx} = \displaystyle\sum_{v=0}^{n-1} \dfrac{(-1)^v a^v x^{n-v}}{(n-v)b^{v+1}} + \dfrac{(-a)^n}{b^{n+1}}\ln|a+bx|$, $n \geq 1$

1.10 $\int \dfrac{x^n dx}{(a+bx)^m} = -\dfrac{x^n}{(m-1)b(a+bx)^{m-1}} + \dfrac{n}{(m-1)b}\int \dfrac{x^{n-1}dx}{(a+bx)^{m-1}}$, $m \geq 2$

1.11 $\int \dfrac{dx}{x(a+bx)} = -\dfrac{1}{a}\ln\left|\dfrac{a}{x}+b\right|$

1.12 $\int \dfrac{dx}{x(a+bx)^2} = \dfrac{1}{a}\left(\dfrac{1}{a+bx} - \dfrac{1}{a}\ln\left|\dfrac{a}{x}+b\right|\right)$

1.13 $\int \dfrac{dx}{x(a+bx)^3} = \dfrac{1}{a}\left[\dfrac{1}{2(a+bx)^2} + \dfrac{1}{a(a+bx)} - \dfrac{1}{a^2}\ln\left|\dfrac{a}{x}+b\right|\right]$

1.14 $\int \dfrac{dx}{x(a+bx)^m} = \sum\limits_{v=1}^{m-1}\dfrac{1}{va^{m-v}(a+bx)^v} - \dfrac{1}{a^m}\ln\left|\dfrac{a}{x}+b\right|$, $m \geq 2$

1.15 $\int \dfrac{dx}{x^2(a+bx)} = -\dfrac{1}{a}\left(\dfrac{1}{x} - \dfrac{b}{a}\ln\left|\dfrac{a}{x}+b\right|\right)$

1.16 $\int \dfrac{dx}{x^2(a+bx)^2} = -\dfrac{1}{a^2}\left(\dfrac{b}{a+bx} + \dfrac{1}{x} - \dfrac{2b}{a}\ln\left|\dfrac{a}{x}+b\right|\right)$

1.17 $\int \dfrac{dx}{x^2(a+bx)^m} = \dfrac{-1}{ax(a+bx)^{m-1}} - \dfrac{mb}{a}\int \dfrac{dx}{x(a+bx)^m}$ (see 14)

1.18 $\int \dfrac{dx}{x^3(a+bx)} = \dfrac{1}{a}\left(\dfrac{b}{ax} - \dfrac{1}{2x^2} - \dfrac{b^2}{a^2}\ln\left|\dfrac{a}{x}+b\right|\right)$

1.19 $\int \dfrac{dx}{x^3(a+bx)^2} = \dfrac{1}{a^2}\left[\dfrac{b^2}{a(a+bx)} + \dfrac{2b}{ax} - \dfrac{1}{2x^2} - \dfrac{3b^2}{a^2}\ln\left|\dfrac{a}{x}+b\right|\right]$

1.20 $\int \dfrac{dx}{x^3(a+bx)^3} = \dfrac{1}{a^3}\left[\dfrac{3b^2}{a(a+bx)} + \dfrac{b^2}{2(a+bx)^2} + \dfrac{3b}{ax} - \dfrac{1}{2x^2} - \dfrac{6b^2}{a^2}\ln\left|\dfrac{a}{x}+b\right|\right]$

2.   Integrals of type $\int x^{\pm n}\dfrac{(a+bx)^{\pm m}dx}{(c+fx)^k}$, $\Delta = af - bc \neq 0$

2.1 $\int \dfrac{a+bx}{c+fx}dx = \dfrac{1}{f}\left(bx + \dfrac{\Delta}{f}\ln|c+fx|\right)$

2.2 $\displaystyle\int\frac{(a+bx)^2}{c+fx}\,dx=\frac{1}{f}\left[\frac{b^2}{2}x^2+b(a+\frac{\Delta}{f})x+\frac{\Delta^2}{f^2}\ln|c+fx|\right]$

2.3 $\displaystyle\int\frac{a+bx}{(c+fx)^2}\,dx=-\frac{1}{f^2}\left[\frac{\Delta}{c+fx}-b\ln|c+fx|\right]$

2.4 $\displaystyle\int\left(\frac{a+bx}{c+fx}\right)^2 dx=\frac{1}{f^2}\left[b^2x-\frac{\Delta}{f}\left(\frac{\Delta}{c+fx}-2b\ln|c+fx|\right)\right]$

2.5 $\displaystyle\int\frac{a+bx}{x(c+fx)}\,dx=\frac{a}{c}\ln|x|-\frac{\Delta}{cf}\ln|c+fx|$

2.6 $\displaystyle\int\frac{dx}{(a+bx)(c+fx)}=-\frac{1}{\Delta}\ln\left|\frac{a+bx}{c+fx}\right|$

2.7 $\displaystyle\int\frac{dx}{(a+bx)(c+fx)^2}=\frac{-1}{\Delta(c+fx)}+\frac{b}{\Delta^2}\ln\left|\frac{a+bx}{c+fx}\right|$

2.8 $\displaystyle\int\frac{dx}{(a+bx)^2(c+fx)^2}=-\frac{1}{\Delta^2}\left(\frac{b}{a+bx}+\frac{f}{c+fx}\right)+\frac{2bf}{\Delta^3}\ln\left|\frac{a+bx}{c+fx}\right|$

2.9 $\displaystyle\int\frac{xdx}{(a+bx)(c+fx)}=\frac{1}{\Delta}\left[\frac{a}{b}\ln|a+bx|-\frac{c}{f}\ln|c+fx|\right]$

2.10 $\displaystyle\int\frac{xdx}{(a+bx)(c+fx)^2}=\frac{c}{f\Delta(c+fx)}-\frac{af}{b\Delta^2}\ln\left|\frac{a+bx}{c+fx}\right|$

3.   Integrals of type $\displaystyle\int\frac{x^{\pm n}dx}{(a^2\pm b^2x^2)^m}$, $a>0$, $b>0$

3.1 $\displaystyle\int\frac{dx}{a^2+b^2x^2}=\frac{1}{ab}\,\text{arctg}\,\frac{bx}{a}$

3.2 $\displaystyle\int\frac{dx}{(a^2+b^2x^2)^2}=\frac{x}{2a^2(a^2+b^2x^2)}+\frac{1}{2a^2b}\,\text{arctg}\,\frac{bx}{a}$

3.3 $\int \dfrac{dx}{(a^2+b^2x^2)^m} = \dfrac{x}{2(m-1)a^2(a^2+b^2x^2)^{m-1}} +$

$+\dfrac{2m-3}{2(m-1)a^2}\int \dfrac{dx}{(a^2+b^2x^2)^{m-1}}$, $m \geq 2$

3.4 $\int \dfrac{xdx}{a^2+b^2x^2} = \dfrac{1}{2b^2}\ln(a^2+b^2x^2)$

3.5 $\int \dfrac{xdx}{(a^2+b^2x^2)^m} = -\dfrac{1}{2(m-1)b^2(a^2+b^2x^2)^{m-1}}$, $m \geq 2$

3.6 $\int \dfrac{x^2dx}{a^2+b^2x^2} = \dfrac{x}{b^2} - \dfrac{a}{b^3}\operatorname{arctg}\dfrac{bx}{a}$

3.7 $\int \dfrac{x^n dx}{(a^2+b^2x^2)^m} = -\dfrac{x^{n-1}}{2(m-1)b^2(a^2+b^2x^2)^{m-1}} +$

$+\dfrac{n-1}{2(m-1)b^2}\int \dfrac{x^{n-2}dx}{(a^2+b^2x^2)^{m-1}}$, $m \geq 2$

3.8 $\int \dfrac{dx}{a^2-b^2x^2} = \dfrac{1}{2ab}\ln\left|\dfrac{a+bx}{a-bx}\right|$

3.9 $\int \dfrac{dx}{(a^2-b^2x^2)^2} = \dfrac{x}{2a^2(a^2-b^2x^2)} + \dfrac{1}{4a^2b}\ln\left|\dfrac{a+bx}{a-bx}\right|$

3.10 $\int \dfrac{dx}{(a^2-b^2x^2)^m} = \dfrac{x}{2(m-1)a^2(a^2-b^2x^2)^{m-1}} +$

$+\dfrac{2m-3}{2(m-1)a^2}\int \dfrac{dx}{(a^2-b^2x^2)^{m-1}}$, $m \geq 2$

3.11 $\int \dfrac{xdx}{a^2-b^2x^2} = -\dfrac{1}{2b^2}\ln|a^2-b^2x^2|$

3.12 $\int \dfrac{xdx}{(a^2-b^2x^2)^m} = \dfrac{1}{2(m-1)b^2(a^2-b^2x^2)^{m-1}}$, $m \geq 2$

3.13 $\int \dfrac{x^2 dx}{a^2 - b^2 x^2} = -\dfrac{x}{b^2} + \dfrac{a}{2b^3} \ln\left|\dfrac{a+bx}{a-bx}\right|$

3.14 $\int \dfrac{x^2 dx}{(a^2 - b^2 x^2)^2} = \dfrac{x}{2b^2(a^2 - b^2 x^2)} - \dfrac{1}{4ab^3} \ln\left|\dfrac{a+bx}{a-bx}\right|$

3.15 $\int \dfrac{x^3 dx}{a^2 - b^2 x^2} = -\dfrac{x^2}{2b^2} - \dfrac{a^2}{2b^4} \ln |a^2 - b^2 x^2|$

3.16 $\int \dfrac{dx}{x(a^2 - b^2 x^2)} = \dfrac{1}{2a^2} \ln\left|\dfrac{x^2}{a^2 - b^2 x^2}\right|$

3.17 $\int \dfrac{dx}{x^2(a^2 - b^2 x^2)} = -\dfrac{1}{a^2 x} + \dfrac{b}{2a^2} \ln\left|\dfrac{a+bx}{a-bx}\right|$

3.18 $\int \dfrac{dx}{x^n (a^2 - b^2 x^2)^m} = -\dfrac{1}{(n-1)a^2 x^{n-1}(a^2 - b^2 x^2)^{m-1}} +$

$+\dfrac{(2m+n-3)b^2}{(n-1)a^2} \int \dfrac{dx}{x^{n-2}(a^2 - b^2 x^2)^m}$, $n \geq 2$

4.   Integrals of type $\int \dfrac{x^{\pm n} dx}{(ax^2 + bx + c)^m}$, $\Delta = b^2 - 4ac \neq 0$

4.1 $\int \dfrac{dx}{ax^2 + bx + c} = \begin{cases} \dfrac{2}{\sqrt{-\Delta}} \operatorname{arctg} \dfrac{2ax+b}{\sqrt{-\Delta}} & , \ \Delta < 0 \\[4mm] \dfrac{1}{\sqrt{\Delta}} \ln\left|\dfrac{2ax+b-\sqrt{\Delta}}{2ax+b+\sqrt{\Delta}}\right| & , \ \Delta > 0 \end{cases}$

4.2 $\int \dfrac{dx}{(ax^2 + bx + c)^2} = \dfrac{-2ax - b}{\Delta(ax^2 + bx + c)} - \dfrac{2a}{\Delta} \int \dfrac{dx}{ax^2 + bx + c}$

4.3 $\int \dfrac{dx}{(ax^2 + bx + c)^m} = \dfrac{-2ax - b}{(m-1)\Delta(ax^2 + bx + c)^{m-1}} -$

$-\dfrac{2(2m-3)}{(m-1)\Delta} \int \dfrac{dx}{(ax^2 + bx + c)^{m-1}}$, $m \geq 2$

4.4 $\displaystyle\int\frac{xdx}{ax^2+bx+c}=$

$$=\begin{cases}\dfrac{1}{2a}\ln|ax^2+bx+c|-\dfrac{b}{a\sqrt{-\Delta}}\arctan\dfrac{2ax+b}{\sqrt{-\Delta}} & ,\Delta<0\\[3mm]\dfrac{1}{2a}\ln|ax^2+bx+c|-\dfrac{b}{2a\sqrt{\Delta}}\ln\left|\dfrac{2ax+b-\sqrt{\Delta}}{2ax+b+\sqrt{\Delta}}\right| & ,\Delta>0\end{cases}$$

4.5 $\displaystyle\int\frac{xdx}{(ax^2+bx+c)^2}=\frac{bx+2c}{\Delta(ax^2+bx+c)}+\frac{b}{\Delta}\int\frac{dx}{ax^2+bx+c}$

4.6 $\displaystyle\int\frac{dx}{x(ax^2+bx+c)}=$

$$=\begin{cases}\dfrac{1}{2c}\ln\dfrac{x^2}{|ax^2+bx+c|}-\dfrac{b}{c\sqrt{-\Delta}}\arctan\dfrac{2ax+b}{\sqrt{-\Delta}} & ,\Delta<0\\[3mm]\dfrac{1}{2c}\ln\dfrac{x^2}{|ax^2+bx+c|}-\dfrac{b}{2c\sqrt{\Delta}}\ln\left|\dfrac{2ax+b-\sqrt{\Delta}}{2ax+b+\sqrt{\Delta}}\right| & ,\Delta>0\end{cases}$$

4.7 $\displaystyle\int\frac{dx}{x(ax^2+bx+c)^2}=\frac{abx-2ac+b^2}{c\Delta(ax^2+bx+c)}+$

$\displaystyle+\frac{1}{2c}\ln\frac{x^2}{|ax^2+bx+c|}++\frac{b(6ac-b^2)}{2c^2\Delta}\int\frac{dx}{ax^2+bx+c}$

5. Integrals of type $\displaystyle\int\frac{x^{\pm n}dx}{\sqrt{(ax^2+bx+c)^m}}$, $m=1,3,5,\ldots$

5.1 $\displaystyle\int\frac{dx}{\sqrt{x^2+px+q}}=\ln|2x+p+2\sqrt{x^2+px+q}\,|$

5.2 $\int \dfrac{dx}{\sqrt{ax^2 + bx + c}} =$

$$= \begin{cases} \dfrac{1}{\sqrt{a}} \ln | 2ax + b + 2\sqrt{a}\sqrt{ax^2 + bx + c} | & , \; a > 0 \; , \; b^2 \neq 4ac \\[3mm] \dfrac{1}{\sqrt{a}} \ln | 2ax + b | & , \; a > 0 \; , \; b^2 = 4ac \\[3mm] -\dfrac{1}{\sqrt{-a}} \arcsin \dfrac{2ax + b}{\sqrt{b^2 - 4ac}} & , \; a < 0 \; , \; b^2 > 4ac \end{cases}$$

5.3 $\int \dfrac{dx}{\sqrt{(ax^2 + bx + c)^3}} = \dfrac{4ax + 2b}{(4ac - b^2)\sqrt{ax^2 + bx + c}}$

5.4 $\int \dfrac{xdx}{\sqrt{ax^2 + bx + c}} = \dfrac{\sqrt{ax^2 + bx + c}}{a} - \dfrac{b}{2a}\int \dfrac{dx}{ax^2 + bx + c}$

5.5 $\int \dfrac{xdx}{\sqrt{(ax^2 + bx + c)^3}} = -\dfrac{2bx + 4c}{(4ac - b^2)\sqrt{ax^2 + bx + c}}$

5.6 $\int \dfrac{x^2 dx}{\sqrt{ax^2 + bx + c}} = \dfrac{2ax - 3b}{4a^2}\sqrt{ax^2 + bx + c} +$

$+\dfrac{3b^2 - 4ac}{8a^2}\int \dfrac{dx}{\sqrt{ax^2 + bx + c}}$

6.   Integrals of type $\int \dfrac{x^{\pm n + 0.5} dx}{(a \pm bx)^m}$, $a > 0$, $b > 0$,

Substitution $\sqrt{x} = t$

6.1 $\int \dfrac{\sqrt{x}dx}{a + bx} = \dfrac{2\sqrt{x}}{b} - \dfrac{2\sqrt{a}}{b\sqrt{b}} \arctan \sqrt{\dfrac{bx}{a}}$

6.2 $\int \dfrac{\sqrt{x}dx}{a - bx} = -\dfrac{2\sqrt{x}}{b} + \dfrac{\sqrt{a}}{b\sqrt{b}} \ln \left| \dfrac{\sqrt{a} + \sqrt{bx}}{\sqrt{a} - \sqrt{bx}} \right|$

6.3 $\displaystyle\int\frac{\sqrt{x}dx}{(a+bx)^2}=-\frac{\sqrt{x}}{b(a+bx)}+\frac{1}{b\sqrt{ab}}\arctan\sqrt{\frac{bx}{a}}$

6.4 $\displaystyle\int\frac{\sqrt{x}dx}{(a-bx)^2}=\frac{\sqrt{x}}{b(a-bx)}-\frac{1}{2b\sqrt{ab}}\ln\left|\frac{\sqrt{a}+\sqrt{bx}}{\sqrt{a}-\sqrt{bx}}\right|$

7.   Integrals of type $\displaystyle\int\frac{x^{\pm n}dx}{\sqrt{(a+bx)^{\pm m}}}$, $m=1,3,5,\ldots$

Substitution $\sqrt{a+bx}=t$

7.1 $\displaystyle\int\frac{dx}{\sqrt{a+bx}}=\frac{2}{b}\sqrt{a+bx}$

7.2 $\displaystyle\int\frac{dx}{\sqrt{(a+bx)^m}}=-\frac{2}{(m-2)b\sqrt{(a+bx)^{m-2}}}$

7.3 $\displaystyle\int\frac{xdx}{\sqrt{(a+bx)^m}}=\frac{2}{b^2\sqrt{(a+bx)^{m-2}}}\left(-\frac{a+bx}{m-4}+\frac{a}{m-2}\right)$

7.4 $\displaystyle\int\frac{x^2dx}{\sqrt{(a+bx)^m}}=$

$\displaystyle=\frac{2}{b^3\sqrt{(a+bx)^{m-2}}}\left[-\frac{(a+bx)^2}{m-6}+\frac{2(a+bx)}{m-4}-\frac{a^2}{m-2}\right]$

7.5 $\displaystyle\int x\sqrt{(a+bx)^m}dx=\frac{2}{b^2}\left[\frac{\sqrt{(a+bx)^{m+4}}}{m+4}-\frac{a\sqrt{(a+bx)^{m+2}}}{m+2}\right]$

7.6 $\displaystyle\int x^n\sqrt{(a+bx)^m}dx=\frac{2\sqrt{(a+bx)^{m+2}}}{b^{n+1}}\sum_{v=0}^{n}\frac{(-1)^v C_n^v(a+bx)^{n-v}a^v}{2n-2v+m+2}$

8.   Integrals of type $\int \dfrac{x^{\pm n} dx}{\sqrt{a^2 + b^2 x^2}}$

8.1 $\quad \displaystyle\int \frac{dx}{\sqrt{a^2 + b^2 x^2}} = \frac{1}{b} \ln | bx + \sqrt{a^2 + b^2 x^2} |$

8.2 $\quad \displaystyle\int \frac{dx}{\sqrt{(a^2 + b^2 x^2)^3}} = \frac{x}{a^2 \sqrt{a^2 + b^2 x^2}}$

8.3 $\quad \displaystyle\int \frac{dx}{\sqrt{(a^2 + b^2 x^2)^5}} = \frac{1}{a^4} \left[\frac{x}{\sqrt{a^2 + b^2 x^2}} - \frac{b^2 x^2}{3 \sqrt{(a^2 + b^2 x^2)^2}} \right]$

8.4 $\quad \displaystyle\int \frac{dx}{\sqrt{(a^2 + b^2 x^2)^m}} =$

$$= \frac{1}{a^{m-1}} \sum_{v=0}^{\frac{m-3}{2}} \frac{(-1)^v C_{\frac{m-3}{2}}^v b^{2v} x^{2v+1}}{(2v+1) \sqrt{(a^2 + b^2 x^2)^{2v+1}}} \quad , \ m = 5, 7, 9, \ldots$$

8.5 $\quad \displaystyle\int \frac{x dx}{\sqrt{a^2 + b^2 x^2}} = \frac{1}{b^2} \sqrt{a^2 + b^2 x^2}$

8.6 $\quad \displaystyle\int \frac{x dx}{\sqrt{(a^2 + b^2 x^2)^2}} = -\frac{1}{b^2 \sqrt{a^2 + b^2 x^2}}$

8.7 $\quad \displaystyle\int \frac{x dx}{\sqrt{(a^2 + b^2 x^2)^m}} = -\frac{1}{(m-2) b^2 \sqrt{(a^2 + b^2 x^2)^{m-2}}} \quad , \ m \geq 3$

8.8 $\quad \displaystyle\int \frac{x^2 dx}{\sqrt{a^2 + b^2 x^2}} = \frac{x \sqrt{a^2 + b^2 x^2}}{2b^2} - \frac{a^2}{2b^2} \ln | bx + \sqrt{a^2 + b^2 x^2} |$

8.9 $\quad \displaystyle\int \frac{x^2 dx}{\sqrt{a^2 + b^2 x^2}} = \frac{\sqrt{(a^2 + b^2 x^2)^3}}{3b^4} - \frac{a^2}{b^4} \sqrt{a^2 + b^2 x^2}$

8.10 $\quad \displaystyle\int \frac{x^3 dx}{\sqrt{(a^2 + b^2 x^2)^3}} = \frac{\sqrt{a^2 + b^2 x^2}}{b^4} + \frac{a^2}{b^4 \sqrt{a^2 + b^2 x^2}}$

8.11 $\int \dfrac{dx}{x\sqrt{a^2+b^2x^2}} = -\dfrac{1}{a}\ln\left|\dfrac{a+\sqrt{a^2+b^2x^2}}{bx}\right|$

8.12 $\int \dfrac{dx}{x^2\sqrt{a^2+b^2x^2}} = -\dfrac{\sqrt{a^2+b^2x^2}}{a^2x}$

8.13 $\int \dfrac{dx}{x^3\sqrt{a^2+b^2x^2}} = -\dfrac{\sqrt{a^2+b^2x^2}}{2a^2x^2} + \dfrac{b^2}{2a^3}\ln\left|\dfrac{a+\sqrt{a+b^2x^2}}{bx}\right|$

8.14 $\int \sqrt{a^2+b^2x^2}\,dx = \dfrac{x\sqrt{a^2+b^2x^2}}{2} + \dfrac{a^2}{2b}\ln\left|bx+\sqrt{a^2+b^2x^2}\right|$

8.15 $\int \sqrt{(a^2+b^2x^2)^3}\,dx =$

$= \dfrac{5a^2x+2b^2x^3}{8}\sqrt{a^2+b^2x^2} + \dfrac{3a^4}{8b}\ln\left|bx+\sqrt{a^2+b^2x^2}\right|$

8.16 $\int x^2\sqrt{a^2+b^2x^2}\,dx =$

$\dfrac{a^2x+2b^2x^2}{8b^2}\sqrt{a^2+b^2x^2} - \dfrac{a^2}{8b^3}\ln\left|bx+\sqrt{a^2+b^2x^2}\right|$

9. Integrals of type $\int \dfrac{x^{\pm n}dx}{\sqrt{a^2-b^2x^2}}$

9.1 $\int \dfrac{dx}{\sqrt{a^2-b^2x^2}} = \dfrac{1}{b}\arcsin\dfrac{bx}{a}$

9.2 $\int \dfrac{dx}{\sqrt{(a^2-b^2x^2)^2}} = \dfrac{x}{a^2\sqrt{a^2-b^2x^2}}$

9.3 $\int \dfrac{xdx}{\sqrt{(a^2-b^2x^2)^m}} =$

$= \dfrac{1}{(m-2)b^2\sqrt{(a^2-b^2x^2)^{m-2}}}$, $m=3,5,7,\dots$

9.4 $\displaystyle\int\frac{x^2 dx}{\sqrt{a^2-b^2x^2}} = -\frac{x\sqrt{a^2-b^2x^2}}{2b^2} + \frac{a^2}{2b^3}\arcsin\frac{bx}{a}$

9.5 $\displaystyle\int\frac{x^2 dx}{\sqrt{(a^2-b^2x^2)^3}} = \frac{x}{b^2\sqrt{a^2-b^2x^2}} - \frac{1}{b^3}\arcsin\frac{bx}{a}$

9.6 $\displaystyle\int\frac{x^2 dx}{\sqrt{(a^2-b^2x^2)^5}} = \frac{x^3}{3a^2\sqrt{(a^2-b^2x^2)^3}}$

9.7 $\displaystyle\int\frac{dx}{x\sqrt{a^2-b^2x^2}} = -\frac{1}{a}\ln\left|\frac{a+\sqrt{a^2-b^2x^2}}{bx}\right|$

9.8 $\displaystyle\int\frac{dx}{x^2\sqrt{a^2-b^2x^2}} = -\frac{\sqrt{a^2-b^2x^2}}{a^2x}$

9.9 $\displaystyle\int\frac{dx}{x^2\sqrt{(a^2-b^2x^2)^3}} = \frac{2b^2x^2-a^2}{a^4x\sqrt{a^2-b^2x^2}}$

9.10 $\displaystyle\int\sqrt{a^2-b^2x^2}\,dx = \frac{x\sqrt{a^2-b^2x^2}}{2} + \frac{a^2}{2b}\arcsin\frac{bx}{a}$

9.11 $\displaystyle\int x\sqrt{a^2-b^2x^2}\,dx = -\frac{\sqrt{(a^2-b^2x^2)^3}}{3b^2}$

9.12 $\displaystyle\int x\sqrt{(a^2-b^2x^2)^m}\,dx = -\frac{\sqrt{(a^2-b^2x^2)^{m+2}}}{(m+2)b^2}$

9.13 $\displaystyle\int\frac{\sqrt{a^2-b^2x^2}}{x}\,dx = \sqrt{a^2-b^2x^2} - a\ln\left|\frac{a+\sqrt{a^2-b^2x^2}}{bx}\right|$

9.14 $\displaystyle\int\frac{\sqrt{(a^2-b^2x^2)^3}}{x}\,dx =$

$$= \frac{4a^2-b^2x^2}{3}\sqrt{a^2-b^2x^2} - a^3\ln\left|\frac{a+\sqrt{a^2-b^2x^2}}{bx}\right|$$

$$9.15 \int \frac{dx}{\sqrt{b^2x^2 - a^2}} = \frac{1}{b} \ln | bx + \sqrt{b^2x^2 - a^2} |$$

$$9.16 \int \frac{xdx}{\sqrt{(b^2x^2 - a^2)^m}} = -\frac{1}{(m-2)b^2\sqrt{(b^2x^2 - a^2)^{m-2}}}$$

$$9.17 \int \frac{x^2dx}{\sqrt{b^2x^2 - a^2}} = \frac{x^2\sqrt{b^2x^2 - a^2}}{2b} + \frac{a^2}{2b^2} \ln | bx + \sqrt{b^2x^2 - a^2} |$$

10.   Integrals of type $\int x^n \sin^m x dx$

$$10.1 \int \sin px \, dx = -\frac{1}{p} \cos px$$

$$10.2 \int \sin^2 px \, dx = \frac{x}{2} - \frac{\sin 2px}{4p}$$

$$10.3 \int \sin^3 px \, dx = \frac{\cos^3 px}{3p} - \frac{\cos px}{p}$$

$$10.4 \int \sin^m px \, dx = -\frac{\sin^{m-1} px \cdot \cos px}{mp} + \frac{m-1}{m} \int \sin^{m-2} px \, dx$$

$$10.5 \int x \sin px \, dx = \frac{1}{p^2} \sin px - \frac{x}{p} \cos px$$

$$10.6 \int x \sin^2 px \, dx = \frac{x^2}{4} - \frac{x \sin 2px}{4} - \frac{\cos 2px}{8p^2}$$

$$10.7 \int x \sin^3 px \, dx = \frac{x \cos 3px}{12p} - \frac{\sin 3px}{36p^2} -$$

$$-\frac{3}{4p} x \cos px + \frac{3}{4p^2} \sin px$$

10.8 $\int x \sin^m px \, dx = \dfrac{\sin^{m-1} px}{m^2 p^2} [\sin px - mpx \cos px] +$

$+ \dfrac{m-1}{m} \int x \sin^{m-2} px \, dx$

10.9 $\int x^2 \sin px \, dx = \dfrac{2x \sin px}{p^2} - \dfrac{p^2 x^2 - 2}{p^3} \cos px$

10.10 $\int x^2 \sin^2 px \, dx = \dfrac{x^3}{6} - \dfrac{2p^2 x^2 - 1}{8p^3} \sin 2px - \dfrac{x \cos 2px}{4p^2}$

10.11 $\int x^n \sin px \, dx = -\dfrac{x^n}{p} \cos px + \dfrac{nx^{n-1}}{p^2} \sin px -$

$- \dfrac{n(n-1)}{p^2} \int x^{n-2} \sin px \, dx$

10.12 $\int x^n \sin^m px \, dx = \dfrac{x^{n-1} \sin^{m-1} px}{m^2 p^2} (n \sin px - mpx \cos px) +$

$+ \dfrac{m-1}{m} \cdot \int x^n \sin^{m-2} px \, dx - \dfrac{n(n-1)}{m^2 p^2} \int x^{n-2} \sin^m px \, dx$

11.   Integrals of type $\int x^{\pm n} \sin^{\pm m} x \, dt$, $\int \dfrac{x^n dx}{(a + b \sin x)^m}$

11.1 $\int \dfrac{dx}{\sin x} = \ln \left| \tan \dfrac{x}{2} \right| = -\dfrac{1}{2} \ln \left| \dfrac{1 + \cos x}{1 - \cos x} \right|$

11.2 $\int \dfrac{dx}{\sin^2 x} = -\cot gx$

11.3 $\int \dfrac{dx}{\sin^3 x} = -\dfrac{\cos x}{2 \sin^2 x} + \dfrac{1}{2} \ln \left| \tan \dfrac{x}{2} \right|$

11.4 $\int \dfrac{dx}{\sin^m px} = -\dfrac{\cos px}{(m-1)p \sin^{m-1} px} + \dfrac{m-2}{m-1} \int \dfrac{dx}{\sin^{m-2} px}$, $m \geq 2$

11.5 $\int \dfrac{x dx}{\sin^2 x} = x \cotg x + \ln |\sin x|$

11.6 $\int \dfrac{dx}{1 \pm \sin x} = \mp \tan\left(\dfrac{\pi}{4} \mp \dfrac{x}{2}\right)$

11.7 $\int \dfrac{dx}{a + b \sin x} =$

$$= \begin{cases} \dfrac{2}{\sqrt{a^2 + b^2}} \arctan \dfrac{a \cdot \tg \dfrac{x}{2} + b}{\sqrt{a^2 - b^2}} & , \ a^2 > b^2 \\[4mm] \dfrac{1}{\sqrt{b^2 - a^2}} \ln \left| \dfrac{a \cdot \tan \dfrac{x}{2} + b - \sqrt{b^2 - a^2}}{a \cdot \tan \dfrac{x}{2} + b + \sqrt{b^2 - a^2}} \right| & , \ a^2 < b^2 \end{cases}$$

11.8 $\int \dfrac{dx}{(1 + \sin x)^2} = -\dfrac{1}{2} \tan\left(\dfrac{\pi}{4} - \dfrac{x}{2}\right) - \dfrac{1}{6} \tan^3\left(\dfrac{\pi}{4} - \dfrac{x}{2}\right)$

11.9 $\int \dfrac{dx}{(1 - \sin x)^2} = \dfrac{1}{2} \cot\left(\dfrac{\pi}{4} - \dfrac{x}{2}\right) + \dfrac{1}{6} \cot^2\left(\dfrac{\pi}{4} - \dfrac{x}{2}\right)$

11.10 $\int \dfrac{dx}{(a + b \sin x)^2} = \dfrac{b \cos x}{(a^2 - b^2)(a + b \sin x)} +$

$+ \dfrac{a}{a^2 - b^2} \int \dfrac{dx}{a + b \sin x}, a^2 \neq b^2$

11.11 $\int \dfrac{x dx}{1 \pm \sin x} = \mp x \cdot \tan\left(\dfrac{\pi}{4} + \dfrac{x}{2}\right) + 2 \ln \left| \dfrac{\cos\left(\dfrac{\pi}{4} \mp \dfrac{x}{2}\right)}{\sin} \right|$

11.12 $\int \dfrac{\sin x dx}{1 \pm \sin x} = \pm x + \tan\left(\dfrac{\pi}{4} \mp \dfrac{x}{2}\right)$

11.13 $\int \dfrac{\sin x dx}{a + b \sin x} = \dfrac{x}{b} - \dfrac{a}{b} \int \dfrac{dx}{a + b \sin x}$, $a^2 \neq b^2$

11.14 $\int \dfrac{\sin x dx}{(1 \pm \sin x)^2} = -\dfrac{1}{2} \tan\left(\dfrac{\pi}{4} \mp \dfrac{x}{2}\right) + \dfrac{1}{6} \tan^3\left(\dfrac{\pi}{4} \mp \dfrac{x}{2}\right)$

11.15 $\int \dfrac{\sin x \, dx}{\sqrt{a^2 - b^2 \sin^2 x}} = -\dfrac{1}{b} \arcsin \dfrac{b \cos x}{\sqrt{a^2 + b^2}}$

11.16 $\int \dfrac{\sin x \, dx}{\sqrt{a^2 - b^2 \sin^2 x}} = -\dfrac{1}{b} \ln |\, b \cos x + \sqrt{a^2 - b^2 \sin^2 x} \,|$

11.17 $\int \dfrac{\sin 2x \, dx}{\sin^n x} = -\dfrac{2}{(n-2)\sin^{n-2} x}$, $n \geq 3$

11.18 $\int \dfrac{\sin x}{\sin 2x} dx = \dfrac{1}{2} \ln \left| \cot\left(\dfrac{x}{2} - \dfrac{\pi}{4} \right) \right|$

11.19 $\int \dfrac{\sin^2 x}{\sin 2x} dx = -\dfrac{1}{2} \ln |\cos x|$

11.20 $\int \dfrac{\sin^3 x}{\sin 2x} dx = -\dfrac{1}{2} \ln \left| \cot\left(\dfrac{x}{2} - \dfrac{\pi}{4} \right) \right| - \dfrac{1}{2} \sin x$

11.21 $\int \dfrac{\sin 3x}{\sin x} dx = x + \sin 2x$

11.22 $\int \dfrac{\sin 3x}{\sin^2 x} dx = 3 \ln \left| \tan \dfrac{x}{2} \right| + 4 \cos x$

11.23 $\int \dfrac{\sin 3x}{\sin^3 x} dx = -3 \cot x - 4x$

12.   Integrals of type $\int x^n \cos px \, dx$, $\int \dfrac{\cos x \, dx}{a + b \cos x}$

12.1 $\int \cos px \, dx = \dfrac{1}{p} \sin px$

12.2 $\int \cos^2 px \, dx = \dfrac{x}{2} + \dfrac{\sin 2px}{4p}$

12.3 $\int \cos^3 px \, dx = \dfrac{1}{p} \sin px - \dfrac{\sin^2 px}{3p}$

12.4 $\int \cos^m px \, dx = \dfrac{\sin px \cos^{m-1} px}{mp} + \dfrac{m-1}{m} \int \cos^{m-2} px \, dx$

12.5 $\int x \cos px \, dx = \dfrac{1}{p^2} \cos px + \dfrac{x}{p} \sin px$

12.6 $\int x \cos^2 px \, dx = \dfrac{x^2}{4} + \dfrac{x \sin 2px}{4p} + \dfrac{\cos 2px}{8p^2}$

12.7 $\int x \cos^3 px \, dx = \dfrac{x \sin 3px}{12p} + \dfrac{\cos 3px}{36p^2} x \sin px + \dfrac{3}{4p^2} \cos px$

12.8 $\int x^2 \cos px \, dx = \dfrac{2x \cos px}{p^2} + \dfrac{p^2 x^2 - 2}{p^2} \sin px$

12.9 $\int x^2 \cos^2 px \, dx = \dfrac{x^3}{6} + \dfrac{2p^2 x^2 - 1}{8p^3} \sin 2px + \dfrac{x \cos 3px}{4p^2}$

12.10 $\int x^3 \cos px \, dx = \dfrac{3p^2 x^2 - 6}{p^4} \cos px + \dfrac{p^2 x^3 - 6x}{p^3} \sin px$

12.11 $\int x^2 \cos^2 px \, dx = \dfrac{x^4}{8} + \dfrac{2p^2 x^3 - 3x}{8p^2} \sin 2px +$

$+ \dfrac{3p^2 x^2 - 3}{16p^4} \cos 2px$

12.12 $\int x^n \cos px \, dx = \dfrac{x^n}{p} \sin px + \dfrac{nx^{n-1}}{p^2} \cos px -$

$- \dfrac{n(n-1)}{p^2} \int x^{n-2} \cos px \, dx$

13. Integrals of type $\int x^{\pm n} \cos^{\pm m} x \, dx$ $\int \dfrac{x^n dx}{(a + b \cos x)^m}$

13.1 $\int \dfrac{dx}{\cos x} = \ln \left| \tan\left(\dfrac{\pi}{4} + \dfrac{x}{2} \right) \right|$

13.2 $\displaystyle\int \frac{dx}{\cos^2 x} = \tan x$

13.3 $\displaystyle\int \frac{dx}{\cos^3 x} = \frac{\sin x}{2\cos^2 x} + \frac{1}{2}\ln\left| \tan\left(\frac{\pi}{4} + \frac{x}{2}\right) \right|$

13.4 $\displaystyle\int \frac{dx}{\cos^m px} = \frac{\sin px}{(m-1)p\cos^{m-1} px} + \frac{m-2}{m-1}\int \frac{dx}{\cos^{m-2} px}$, $m \geq 2$

13.5 $\displaystyle\int \frac{xdx}{\cos^2 x} = x\tan x + \ln|\cos x|$

13.6 $\displaystyle\int \frac{xdx}{\cos^3 x} = \frac{x\sin x}{2\cos^2 x} - \frac{1}{2\cos x} + \frac{1}{2}\int \frac{xdx}{\cos x}$

13.7 $\displaystyle\int \frac{dx}{a + b\cos x} =$

$$= \begin{cases} \dfrac{2}{\sqrt{a^2 - b^2}} \arctan \dfrac{(a-b)\tan\dfrac{x}{2}}{\sqrt{a^2 - b^2}} & , a^2 > b^2 \\[4ex] \dfrac{1}{\sqrt{b^2 - a^2}} \ln\left| \dfrac{(b-a)\tan\dfrac{x}{2} + \sqrt{b^2 - a^2}}{(b-a)\tan\dfrac{x}{2} - \sqrt{b^2 - a^2}} \right| & , a^2 < b^2 \end{cases}$$

13.8 $\displaystyle\int \frac{dx}{1 \pm \cos x} = \pm\tan\left[\frac{\pi}{4} \mp \left(\frac{\pi}{4} - \frac{x}{2}\right)\right]$

13.9 $\displaystyle\int \frac{dx}{(1 + \cos x)^2} = \frac{1}{2}\cot\frac{x}{2} + \frac{1}{6}\tan^3\frac{x}{2}$

13.10 $\displaystyle\int \frac{dx}{(1 - \cos x)^2} = -\frac{1}{2}\cot\frac{x}{2} - \frac{1}{6}\cot^3\frac{x}{2}$

13.11 $\displaystyle\int \frac{xdx}{1 \pm \cos x} = \pm x\tan\left[\frac{\pi}{4} \mp \left(\frac{\pi}{4} - \frac{x}{2}\right)\right] + 2\ln\left|\frac{\pi}{4} \mp \left(\frac{\pi}{4} - \frac{x}{2}\right)\right|$

13.12 $\int \dfrac{\cos x \, dx}{1 \pm \cos x} = \pm x \mp \tan\left[\dfrac{\pi}{4} \mp \left(\dfrac{\pi}{4} - \dfrac{x}{2}\right)\right]$

13.13 $\int \cos px \cdot \cos qx \, dx = \dfrac{\sin(p+q)x}{2(p+q)} + \dfrac{\sin(p-q)x}{2(p-q)}$, $p^2 \neq q^2$

13.14 $\int \dfrac{\cos x \, dx}{\sqrt{a^2 - b^2 \cos^2 x}} = \dfrac{1}{b} \ln | b \sin x + \sqrt{a^2 - b^2 \cos^2 x} |$

13.15 $\int \dfrac{\cos 2x}{\cos x} \, dx = 2 \sin x - \ln \left| \tan\left(\dfrac{\pi}{4} + \dfrac{x}{2}\right) \right|$

13.16 $\int \dfrac{\cos x}{\cos 2x} \, dx = \dfrac{1}{2\sqrt{2}} \ln \left| \dfrac{1 - \sqrt{2} \sin x}{1 + \sqrt{2} \sin x} \right|$

13.17 $\int \dfrac{\cos^2 x \, dx}{\cos 2x} = \dfrac{x}{2} - \dfrac{1}{4} \ln \left| \dfrac{1 - \tan x}{1 + \tan x} \right|$

13.18 $\int \dfrac{\cos 3x}{\cos x} \, dx = \sin 2x - x$

13.19 $\int \dfrac{\cos 3x}{\cos^2 x} \, dx = 4 \sin x - 3 \ln \left| \tan\left(\dfrac{\pi}{4} + \dfrac{x}{2}\right) \right|$

14.   Integrals of type $\int \sin^{\pm m} x \cdot \cos^{\pm n} x \, dx$

14.1 $\int \sin x \cos x \, dx = \dfrac{\sin^2 x}{2}$

14.2 $\int \sin x \cos^n x \, dx = -\dfrac{\cos^{n+1} x}{n+1}$

14.3 $\int \sin^m x \cos x \, dx = \dfrac{\sin^{m+1} x}{m+1}$

14.4 $\int \sin^2 x \cos^2 x \, dx = \dfrac{1}{8}\left(x - \dfrac{\sin 4x}{4}\right)$

14.5 $\int \sin^2 x \cos^3 x \, dx = \dfrac{\sin^3 x \cos^2 x}{5} + \dfrac{2}{15} \sin^3 x$

14.6 $\int \sin^m x \cos^n x \, dx =$

$$= \frac{\sin^{m+1} x \cdot \cos^{n-1} x}{m+n} + \frac{n-1}{m+n} \cdot \int \sin^m x \cos^{n-2} x \, dx$$

14.7 $\int \dfrac{dx}{\sin x \cos x} = \ln |\tan x|$

14.8 $\int \dfrac{dx}{\sin x \cos^n x} = \dfrac{1}{(n-1)\cos^{n-1} x} + \int \dfrac{dx}{\sin x \cos^{n-2} x}$, $n \geq 2$

14.9 $\int \dfrac{dx}{\sin^m x \cos x} = -\dfrac{1}{(m-1)\sin^{m-1} x} + \int \dfrac{dx}{\sin^{m-2} x \cos x}$, $m \geq 2$

14.10 $\int \dfrac{\sin 2x}{\cos x} \, dx = -2 \cos x$

14.11 $\int \dfrac{\sin 2x}{\cos^2 x} \, dx = -2 \ln |\cos x|$

14.12 $\int \dfrac{\cos 2x}{\sin x} \, dx = 2 \cos x + \ln \left| \operatorname{tg} \dfrac{x}{2} \right|$

14.13 $\int \dfrac{\cos 2x}{\sin^2 x} \, dx = -\cot x - 2x$

14.14 $\int \dfrac{\sin px}{\cos^n x} \, dx = 2 \int \dfrac{\sin(p-1)x}{\cos^{n-1} x} \, dx - \int \dfrac{\sin(p-2)x}{\cos^n x} \, dx$

15.   Integrals of type $\int \dfrac{\sin x \, dx}{a + b \cos x}$, $\int \dfrac{\cos x \, dx}{a + b \sin x}$

15.1 $\int \dfrac{\sin x \, dx}{a + b \cos x} = -\dfrac{1}{b} \ln |a + b \cos x|$

15.2 $\int \dfrac{\sin x \, dx}{\cos x (1 \pm \cos x)} = \ln \left| \dfrac{1 \pm \cos x}{\cos x} \right|$

15.3 $\displaystyle\int\frac{\sin x\,dx}{\cos x\,(1\pm\sin x)}=\frac{1}{2(1\pm\sin x)}\pm\frac{1}{2}\ln\left|\tan\left(\frac{\pi}{4}+\frac{x}{2}\right)\right|$

15.4 $\displaystyle\int\frac{\sin x\,dx}{(a+b\cos x)(\alpha+\beta\cos x)}=\frac{1}{a\beta-b\alpha}\ln\left|\frac{a+b\cos x}{\alpha+\beta\cos x}\right|$

15.5 $\displaystyle\int\frac{\cos x\,dx}{a+b\sin x}=\pm\ln|a+b\sin x|$

15.6 $\displaystyle\int\frac{\cos x\,dx}{\sin x\,(1\pm\sin x)}=\ln\left|\frac{\sin x}{1\pm\sin x}\right|$

15.7 $\displaystyle\int\frac{\cos x\,dx}{\sin x\,(1\pm\cos x)}=-\frac{1}{2(1\pm\cos x)}\pm\frac{1}{2}\ln\left|\tan\frac{x}{2}\right|$

15.8 $\displaystyle\int\frac{\cos x\,dx}{(a+b\sin x)(\alpha+\beta\sin x)}=\frac{1}{a\beta-b\alpha}\ln\left|\frac{\alpha+\beta\sin x}{a+b\sin x}\right|$

15.9 $\displaystyle\int\frac{dx}{\sin x\,(1\pm\cos x)}=\pm\frac{1}{2(1\pm\cos x)}+\frac{1}{2}\ln\left|\tan\frac{x}{2}\right|$

15.10 $\displaystyle\int\frac{dx}{\cos x\,(1\pm\sin x)}=\mp\frac{1}{2(1\pm\sin x)}+\frac{1}{2}\ln\left|\tan\left(\frac{x}{2}+\frac{\pi}{4}\right)\right|$

15.11 $\displaystyle\int\frac{dx}{\sin x\pm\cos x}=\frac{1}{\sqrt{2}}\ln\left|\tan\left(\frac{x}{2}\pm\frac{\pi}{8}\right)\right|$

15.12 $\displaystyle\int\frac{dx}{a\cos x+b\sin x}=\frac{1}{\sqrt{a^2+b^2}}\ln\left|\tan\frac{x+\arccos\dfrac{b}{\sqrt{a^2+b^2}}}{2}\right|$

15.13 $\displaystyle\int R(\sin x,\cos x)\,dx=2\int R\left(\frac{2t}{1+t^2},\frac{1-t^2}{1+t^2}\right)\frac{t\,dt}{1+t^2}$

Substitution $t = \tan\dfrac{x}{2}$

16.   Integrals with $\tan x$ and $\cot x$

16.1 $\displaystyle\int \tan x\,dx = \ln|\cos x|$

16.2 $\displaystyle\int \tan^2 x\,dx = \tan x - x$

16.3 $\displaystyle\int \tan^3 x\,dx = \dfrac{\tan^2 x}{2} + \ln|\cos x|$

16.4 $\displaystyle\int \tan^n x\,dx = \dfrac{\tan^{n-1} x}{n-1} - \int \tan^{n-2} x\,dx \ , \ n \geq 2$

16.5 $\displaystyle\int \cot x\,dx = \ln|\sin x|$

16.6 $\displaystyle\int \cot^2 x\,dx = -\cot x - x$

16.7 $\displaystyle\int \cot^3 x\,dx = -\dfrac{\cot^2 x}{2} - \ln|\sin x|$

16.8 $\displaystyle\int \cot^n x\,dx = -\dfrac{\cot^{n-1} x}{n-1} - \int \cot^{n-2} x\,dx \ , \ (n \geq 2)$

16.9 $\displaystyle\int \dfrac{dx}{\tan x \pm 1} = \pm\dfrac{x}{2} + \dfrac{1}{2}\ln|\sin x \pm \cos x|$

16.10 $\displaystyle\int \dfrac{dx}{a + b\tan x} = \dfrac{1}{a^2 + b^2}(b\ln|a + b\tan x| + b\ln|\cos x| + ax)$

16.11 $\displaystyle\int \dfrac{\tan x\,dx}{\tan x \pm 1} = \dfrac{x}{2} \mp \dfrac{1}{2}\ln|\sin x \pm \cos x|$

16.12 $\displaystyle\int \dfrac{\tan x\,dx}{a + b\tan x} = \dfrac{1}{a^2 + b^2}(bx - a\ln|a\cos x + b\sin x|)$

16.13 $\displaystyle\int \dfrac{dx}{1 + \tan^2 x} = \dfrac{x}{2} + \dfrac{1}{4}\sin 2x$

16.14 $\int \dfrac{dx}{a^2 + b^2 \tan^2 x} = \dfrac{1}{a^2 + b^2}\left[x - \left|\dfrac{b}{a}\right| \arctan\left(\left|\dfrac{b}{a}\right| \operatorname{tg} x\right)\right]$, $a^2 \neq b^2$

17. Integrals with $\arcsin x$, $\arccos x$, $\arctan x$, $\operatorname{arccot} x$

17.1 $\int \arcsin \dfrac{x}{a}\, dx = x \arcsin \dfrac{x}{a} + \sqrt{a^2 - x^2}$

17.2 $\int \left(\arcsin \dfrac{x}{a}\right)^2 dx = x\left(\arcsin \dfrac{x}{a}\right)^2 - 2x + 2\sqrt{a^2 - x^2}\, \arcsin \dfrac{x}{a}$

17.3 $\int \left(\arcsin \dfrac{x}{a}\right)^n dx = a\int t^n \cos t\, dt$, **Substitution** $t = \arcsin \dfrac{x}{a}$

17.4 $\int x \arcsin \dfrac{x}{a}\, dx = \dfrac{2x^2 - a^2}{4} \arcsin \dfrac{x}{a} + \dfrac{x}{4}\sqrt{a^2 - x^2}$

17.5 $\int x^n \arcsin \dfrac{x}{a}\, dx = \dfrac{x^{n+1}}{n+1} \arcsin \dfrac{x}{a} - \dfrac{1}{n+1}\int \dfrac{x^{n+1}}{\sqrt{a^2 - x^2}}\, dx$

17.6 $\int \arccos \dfrac{x}{a}\, dx = x \arccos \dfrac{x}{a} - \sqrt{a^2 - x^2}$

17.7 $\int x \arccos \dfrac{x}{a}\, dx = \dfrac{2x^2 - a^2}{4} \arccos \dfrac{x}{a} - \dfrac{x}{4}\sqrt{a^2 - x^2}$

17.8 $\int x^2 \arccos \dfrac{x}{a}\, dx = \dfrac{x^2}{3} \arccos \dfrac{x}{a} - \dfrac{1}{9}(x^2 + 2a^2)\sqrt{a^2 - x^2}$

17.9 $\int \arctan \dfrac{x}{a}\, dx = x \arctan \dfrac{x}{a} - \dfrac{a}{2} \ln(a^2 + x^2)$

17.10 $\int \left(\arctan \dfrac{x}{a}\right)^2 dx = x\left(\arctan \dfrac{x}{a}\right)^2 - 2a\int \dfrac{x \arctan \dfrac{x}{a}}{a^2 + x^2}\, dx$

17.11 $\int x \arctan \dfrac{x}{a}\, dx = \dfrac{1}{2}(x^2 + a^2) \arctan \dfrac{x}{a} - \dfrac{ax}{2}$

17.12 $\int x^2 \arctan \dfrac{x}{a} dx = \dfrac{x^3}{3} \arctan \dfrac{x}{a} - \dfrac{ax^2}{6} + \dfrac{a^3}{6} \ln(a^2 + x^2)$

17.13 $\int x^n \arctan \dfrac{x}{a} dx = \dfrac{x^{n+1}}{n+1} \arctan \dfrac{x}{a} - \dfrac{a}{n+1} \int \dfrac{x^{n+1}}{a^2 + x^2} dx$

17.14 $\int \dfrac{1}{x^2} \arctan \dfrac{x}{a} = -\dfrac{1}{x} \arctan \dfrac{x}{a} - \dfrac{1}{2a} \ln \dfrac{a^2 + x^2}{x^2}$

18.   Integrals with e^{ax}

18.1 $\int A^{ax+b} dx = \dfrac{1}{a \ln A} A^{ax+b}$, $A > 0$, $A \neq 1$

18.2 $\int xe^{ax} dx = \dfrac{ax-1}{a^2} e^{ax}$

18.3 $\int x^2 e^{ax} dx = \dfrac{a^2 x^2 - 2ax + 2}{a^3} e^{ax}$

18.4 $\int x^3 e^{ax} dx = \dfrac{a^3 x^3 - 3a^2 x^2 + 6ax - 6}{a^4} e^{ax}$

18.5 $\int x^n e^{ax} dx = \dfrac{x^n e^{ax}}{a} - \dfrac{n}{a} \int x^{n-1} e^{ax} dx$

18.6 $\int \dfrac{dx}{\alpha + \beta e^x} = \dfrac{x}{\alpha} - \dfrac{1}{\alpha} \ln|\alpha + \beta e^x|$

18.7 $\int \dfrac{dx}{(\alpha + \beta e^x)^2} = -\dfrac{1}{\alpha^2} \left[\ln \left| \dfrac{\alpha + \beta e^x}{e^x} \right| + \dfrac{\beta e^x}{\alpha + \beta e^x} \right]$

18.8 $\int \dfrac{e^{2ax} dx}{\alpha + \beta e^{ax}} = \dfrac{e^{ax}}{a\beta} - \dfrac{\alpha}{a\beta^2} \ln|\alpha + \beta e^{ax}|$

18.9 $\int \dfrac{dx}{\sqrt{e^x - 1}} = 2 \operatorname{arctg} \sqrt{e^x - 1}$

18.10 $\int \frac{e^x dx}{\sqrt{1 \pm e^x}} = \pm 2\sqrt{1 \pm e^x}$, **Substitution** $e^x = t$

18.11 $\int \frac{e^{ax} dx}{\sqrt{(\alpha \pm \beta e^{ax})^m}} = \mp \frac{2}{a\beta(m-2)\sqrt{(\alpha \pm \beta e^{ax})^{m-2}}}$, **Substitution**

$\alpha \pm \beta e^{ax} = t$

18.12 $\int e^{ax} \sin px \, dx = \frac{e^{ax}(a\sin px - p\cos px)}{a^2 + p^2}$

18.13 $\int e^{ax} \cos p \, x dx = \frac{e^{ax}(a\cos px + p\sin px)}{a^2 + p^2}$

Denotations

$\mathbb{N}$ - nartural numbers $\mathbb{C}$ - complex numbers

$\mathbb{Z}$ - integers   **a,u,v** - vectors

$\mathbb{Q}$ - rational numbers α, a, b - scalars

$\mathbb{R}$ - real numbers   n! - n factorial

The page number refers to the first page in which the sign is used.

Index

B

Basis

orthogonal 202

orthogonalization...201

Benomial (power) expansion..120

Benomial series 120

Bernoulli's equation...209

Bessel equation...219

functions of first kind 219

Bessel inequality 203,248

Binomial coefficients 281

Block diagonal matrix 196

C

Cauchy Schwartz inequality......................................95,200

Chain rule 68,132

Characteristic equation 212,213,218

Characteristic polynomial..194

matrix 194

Chebyshev's equation 217

Circle 26

tangent line 26

Class C 129

Class C^n 129

Class of integrable functions 93

Cofactor..180,181

Combination...281

Complex functions 225

branch point..229

continuous 230

derivative..233

Cauchy-Riemann criterion 234

elementary functions 230,232

Euler formula 231

I

L

Matrix(ces)

T

W